SUMMIT MATH

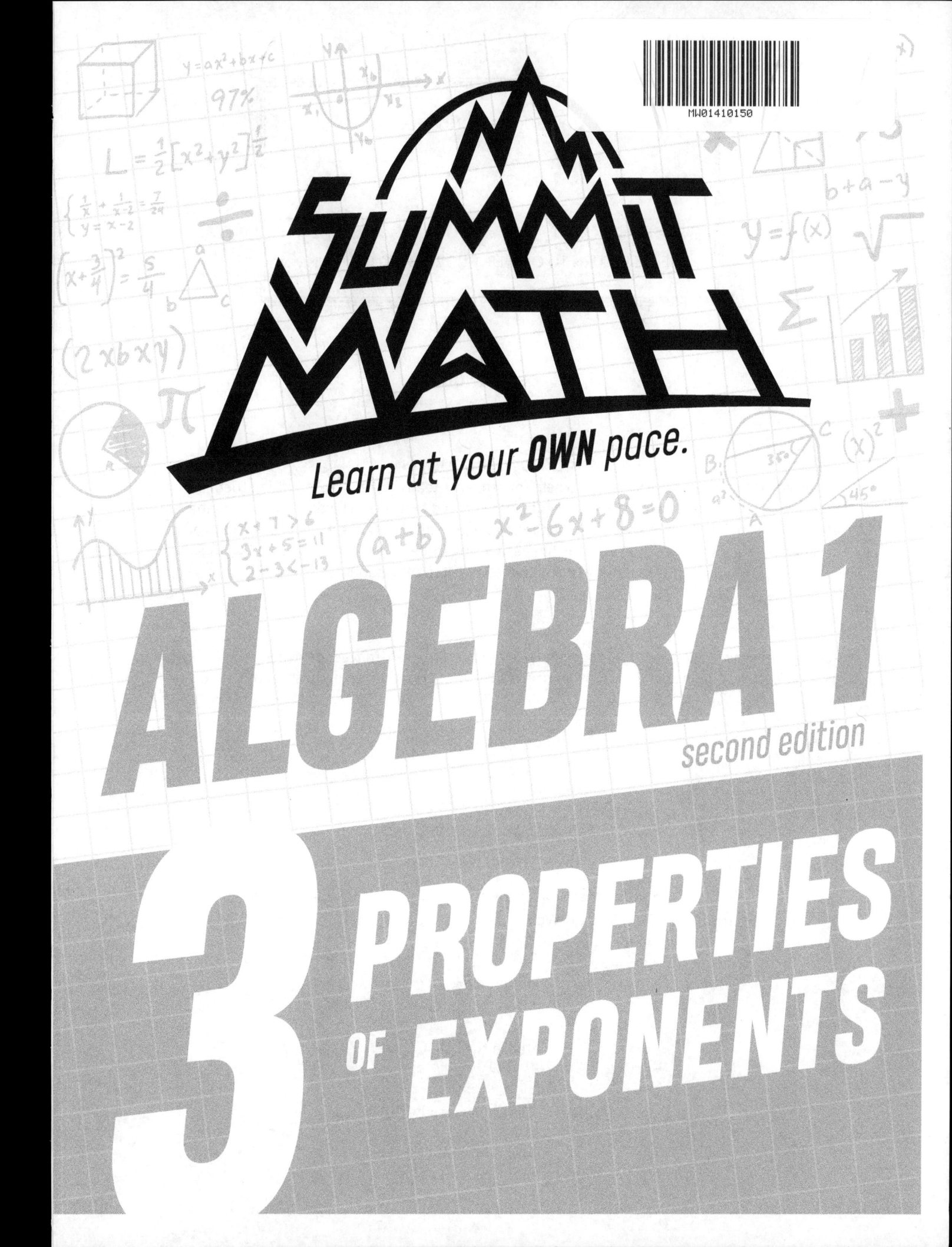

*Learn at your **OWN** pace.*

ALGEBRA 1

second edition

3 PROPERTIES OF EXPONENTS

```
┌─────────────────────────────────────────────┐
│                                             │
│           THIS BOOK BELONGS TO:             │
│                                             │
│                                             │
│   _____  │
│                                             │
│   PLEASE RETURN TO OWNER IF BORROWED OR FOUND │
│                                             │
└─────────────────────────────────────────────┘
```

DEDICATION
To Lauren, Chloe, Dawson and Teagan

ACKNOWLEDGEMENTS
I started writing these books in 2013 to help my students learn better. I kept writing them because I received encouraging feedback from students, parents and teachers. Thank you to all who have used these books, pointed out my mistakes, and made suggestions along the way. Thank you to all of the students and parents who asked me to keep writing more books. Thank you to my family for supporting me through every step of this journey.

All rights reserved. No part of this book may be reproduced, transmitted, or stored in an information retrieval system in any form or by any means without prior written permission of the author.

Copyright © 2020

This book was typeset in the following fonts:
Seravek + Mohave + *Heading Pro*

Graphics in Summit Math books are made using the following resources:
Microsoft Excel | Microsoft Word | Desmos | Geogebra | Adobe Illustrator

First printed in 2017

Printed in the U.S.A.

Summit Math Books are written by Alex Joujan.

www.summitmathbooks.com

INTRODUCTION

Learning math through Guided Discovery:
A Guided Discovery learning experience is designed to help you experience a feeling of discovery as you learn each new topic.

Why this curriculum series is named Summit Math:
Learning through Guided Discovery can be compared to climbing a mountain. Climbing and learning both require effort and persistence. In both activities, people naturally move at different paces, but they can reach the summit if they keep moving forward. Whether you race rapidly through these books or step slowly through each scenario, this curriculum is designed to keep advancing your learning until you reach the end of the book.

Guided Discovery Scenarios:
The Guided Discovery Scenarios in this book are written and arranged to show you that new math concepts are related to previous concepts you have already learned. Try to fully understand each scenario before moving on to the next one. To do this, try the scenario on your own first, check your answer when you finish, and then fix any mistakes, if needed. Making mistakes and struggling are essential parts of the learning process.

Homework and Extra Practice Scenarios:
After you complete the scenarios in each Guided Discovery section, you may think you know those topics well, but over time, you will forget what you have learned. Extra practice will help you develop better retention of each topic. Use the Homework and Extra Practice Scenarios to improve your understanding and to increase your ability to retain what you have learned.

The Answer Key:
The Answer Key is included to promote learning. When you finish a scenario, you can get immediate feedback. When the Answer Key is not enough to help you fully understand a scenario, you should try to get additional guidance from another student or a teacher.

Star symbols:
Scenarios marked with a star symbol ★ can be used to provide you with additional challenges. Star scenarios are like detours on a hiking trail. They take more time, but you may enjoy the experience. If you skip scenarios marked with a star, you will still learn the core concepts of the book.

To learn more about Summit Math and to see more resources:
Visit www.summitmathbooks.com.

GUIDED DISCOVERY SCENARIOS

As you complete scenarios in this part of the book, follow the steps below.

Step 1: Try the scenario.
Read through the scenario on your own or with other classmates. Examine the information carefully. Try to use what you already know to complete the scenario. Be willing to struggle.

Step 2: Check the Answer Key.
When you look at the Answer Key, it will help you see if you fully understand the math concepts involved in that scenario. It may teach you something new. It may show you that you need guidance from someone else.

Step 3: Fix your mistakes, if needed.
If there is something in the scenario that you do not fully understand, do something to help you understand it better. Go back through your work and try to find and fix your errors. Mistakes provide an opportunity to learn. If you need extra guidance, get help from another student or a teacher.

After Step 3, go to the next scenario and repeat this 3-step cycle.

NEED EXTRA HELP?
watch videos online

Teaching videos for every scenario in the Guided Discovery section of this book are available at www.summitmathbooks.com/algebra-1-videos.

GUIDED DISCOVERY SCENARIOS

CONTENTS

Section 1 **INTRODUCTION TO EXPONENTS** **3**

Section 2 **MULTIPLYING EXPRESSIONS WITH EXPONENTS** **7**

Section 3 **DIVIDING EXPRESSIONS WITH EXPONENTS** **11**

Section 4 **RAISING AN EXPONENT TO AN EXPONENT** **15**

Section 5 **EXPONENTS REVIEW** **19**

Section 6 **THE EXPONENT OF ZERO** **23**

Section 7 **NEGATIVE EXPONENTS** **25**

Section 8 **EVALUATING EXPRESSIONS WITH EXPONENTS** **31**

Section 9 **CUMULATIVE REVIEW** **35**

Section 10 **ANSWER KEY** **39**

HOMEWORK & EXTRA PRACTICE SCENARIOS **43**

Section 1
INTRODUCTION TO EXPONENTS

GUIDED DISCOVERY SCENARIOS

1. How many squares do you see in the image below?

   [grid of squares]

   Counting the squares above is an easier task for an older student than it is for, say, a five-year-old child. To a kindergartener, the likely way to approach the task above is by counting... one square at a time. 1, 2, 3, 4, ..., 27, 28, 29, 30, ... and so on.

   Fortunately, you don't need to count like this any more. At some point, you learned how to speed up the tedious task of counting by combining groups of numbers and counting those groups multiple times: hence the word <u>multiplication</u>. For example, "4 + 4 + 4 + 4 + 4 + 4 + 4 + 4" can be expressed as "8 sets of 4" or "8 times 4". Multiplication is a way to represent repeated addition.

2. Consider an example of another repeated operation. You'll need a spare sheet of paper.

    a. Take a sheet of paper and fold the paper in half once. When you unfold it, the paper will now be divided into how many sections?

    b. Fold the paper in half once again. Now fold that portion in half, and then fold this entire portion in half a third time. When you open it up, the paper will be divided into how many sections?

    c. If you fold a sheet of paper in half 5 times and open it up again, into how many sections will the paper be divided?

3. You should now notice a pattern in the growth in the number of sections as the paper is folded.

    a. A sheet of paper will be divided into 64 sections if you fold it in half ____ times.

    b. A sheet of paper will be divided into 256 sections if you fold it in half ____ times.

4. If you fold a sheet of paper in half N times and open it up again, into how many sections will the paper be divided?

GUIDED DISCOVERY SCENARIOS

If you could fold a sheet of paper in half 10 times and open it up again, it would be divided into $2 \cdot 2 \cdot 2 \cdot 2 \cdot 2 \cdot 2 \cdot 2 \cdot 2 \cdot 2 \cdot 2$ sections. If you could fold it in half 75 times and open it up again, it would be divided into... as with the squares earlier, this is getting rather tedious again without more condensed notation.

5. Since "3 + 3 + 3 + 3 + 3 + 3 + 3" can be expressed in a more concise form as "3×7", how can you express $2 \cdot 2 \cdot 2 \cdot 2 \cdot 2 \cdot 2 \cdot 2 \cdot 2 \cdot 2 \cdot 2$ in a more concise form?

Just as multiplication represents repeated addition, an exponent is a type of notation that is used to represent repeated multiplication. An exponent can be used to write $2 \cdot 2 \cdot 2 \cdot 2 \cdot 2 \cdot 2 \cdot 2 \cdot 2 \cdot 2 \cdot 2$ in a much more condensed form as 2^{10} (2 raised to the 10th power).

6. Write each of the following expressions as a single number raised to an exponent.

    a. $5 \cdot 5 \cdot 5$
    b. $10 \cdot 10 \cdot 10 \cdot 10$
    c. $(-3) \cdot (-3) \cdot (-3) \cdot (-3) \cdot (-3)$

7. Write each of the following expressions as a single variable raised to an exponent.

    a. $x \cdot x \cdot x$
    b. $y \cdot y \cdot y \cdot y$
    c. $z \cdot z \cdot z \cdot z \cdot z$

8. Write each of the following expressions as a repeated multiplication expression. For example, 7^3 can be written as $7 \cdot 7 \cdot 7$.

    a. 10^3
    b. $(-2)^4$
    c. m^5

9. Write each of the following expressions as a repeated multiplication expression. For example, 7^3 can be written as $7 \cdot 7 \cdot 7$.

    a. $3^3 \cdot 3^2$
    b. $y^3 \cdot y^2 \cdot y$

10. How would you write $(2x)^4$ as a repeated multiplication expression?

11. How would you write $(x+1)^2$ as a repeated multiplication expression?

NOTES

Use this page to record important ideas in the previous section or for any other writing that helps you learn the topics in this book.

Section 2
MULTIPLYING EXPRESSIONS WITH EXPONENTS

GUIDED DISCOVERY SCENARIOS

12. Consider the result when expressions with exponents are multiplied together. Simplify each product below and write each result as a single number raised to an exponent.

    a. $3^3 \cdot 3^2$
    b. $4^2 \cdot 4^6$
    c. $5^3 \cdot 5^6$
    d. $6^{10} \cdot 6^{10}$

13. Write each expression as a single number raised to an exponent.

    a. $3^2 \cdot 3 \cdot 3^4$
    b. $(-1)^7 \cdot (-1)^2 \cdot (-1)$
    c. $2^x \cdot 2^y$

14. Write each expression as a single variable raised to an exponent.

    a. $x^2 \cdot x$
    b. $x^2 \cdot x^3$
    c. $x^3 \cdot x^5$

15. Write each expression as a single variable raised to an exponent.

    a. $x^2 \cdot x \cdot x^4$
    b. $y^7 \cdot y^2 \cdot y$
    c. $x^A \cdot x^B \cdot x^C$

16. The previous scenarios illustrate The Product Rule, which applies when you multiply like bases and determine the exponent of your result. Write out The Product Rule rule in your own words as if you were explaining it to someone.

17. Consider that $5 \cdot 5 = 25$.

    a. Does $5^2 \cdot 5^3$ have the same value as 25^5 or does $5^2 \cdot 5^3$ equal 5^5? How can you persuade someone else that your conclusion is accurate?

    b. Does $3^5 \cdot 3^4$ have the same value as 3^9 or does $3^5 \cdot 3^4$ equal 9^9?

GUIDED DISCOVERY SCENARIOS

18. Simplify each expression to make each base appear only once.

    a. $10 \cdot 10 \cdot x \cdot x$

    b. $4y \cdot 4y \cdot 4$

19. Simplify each expression to make each base appear only once.

    a. $2^5 \cdot 2 \cdot 6^7 \cdot 6^3$

    b. $7x^2 \cdot 7x^2 \cdot 7x^2$

    c. $p \cdot p^2 \cdot m^6 \cdot m^3 \cdot p$

20. Simplify each expression to make each base appear only once.

    a. $3^x \cdot 3^x$

    b. $8^y \cdot 8^y \cdot 8^5$

    c. $5^x \cdot 5 \cdot 5^{2x}$

21. Fill in the blank.

    a. $3x^3 \cdot 5x^2 = \underline{} x^5$

    b. $x^7 \cdot 3x^3 = 3\underline{}$

    c. $-2x \cdot -x^6 = \underline{}$

22. Simplify each expression as much as you can.

    a. $-x^2 \cdot 4x$

    b. $3y^6 \cdot 7y^4$

    c. $\left(-9x^3\right) \cdot \left(-5x^5\right) \cdot x$

23. Simplify each expression.

    a. $\left(3w^5\right)^2$

    b. $2x^{10} \cdot 9x^{-4}$

    c. $\left(-4m^7\right)^2$

NOTES

Use this page to record important ideas in the previous section or for any other writing that helps you learn the topics in this book.

Section 3
DIVIDING EXPRESSIONS WITH EXPONENTS

GUIDED DISCOVERY SCENARIOS

24. Consider the fraction $\frac{10}{5}$. If you write it as $\frac{2 \cdot 5}{5}$, or $\frac{2}{1} \cdot \frac{5}{5}$, you can see a disguised form of 1 expressed as the fraction $\frac{5}{5}$. When a number is multiplied by 1, its value does not change. This is one way to see why the fraction $\frac{10}{5}$ can be simplified and written as _____.

25. How many times does a disguised form of 1 appear in the expression $2 \cdot 2 \cdot 2 \cdot \frac{2}{2} \cdot \frac{2}{2}$?

26. Write the expression in the previous scenario as a single "2" raised to an exponent.

27. How many times does a disguised form of 1 appear in each expression below?

    a. $\frac{3 \cdot 3 \cdot 3 \cdot 3}{3 \cdot 3}$
    b. $\frac{7 \cdot 7 \cdot 7 \cdot 7 \cdot 7}{7 \cdot 7 \cdot 7}$
    c. $\frac{4 \cdot 4 \cdot 4 \cdot 4 \cdot 4 \cdot 4 \cdot 4 \cdot 4}{4^6}$

28. Write each expression in the previous scenario as a single number raised to an exponent.

29. Write each expression as a single number (a base) raised to an exponent.

    a. $\frac{3^4}{3^2}$
    b. $\frac{7^5}{7^3}$
    c. $\frac{9^2}{9^1}$

30. Write each expression as a single base raised to an exponent.

    a. $\frac{2^6}{2^3}$
    b. $\frac{4^{10}}{4^9}$
    c. $\frac{5^{11}}{5^{11}}$

31. Write each expression as a base raised to an exponent.

    a. $5^7 \div 5^4$
    b. $(-1)^7 \div (-1)^4$
    c. $x^7 \div x^4$

GUIDED DISCOVERY SCENARIOS

32. Write each expression as a base raised to an exponent.

   a. $\dfrac{x^2}{x}$ b. $\dfrac{x^5}{x^2}$ c. $\dfrac{x^4}{x^4}$ d. $\dfrac{x^{10}}{x^2}$ e. $\dfrac{x^A}{x^B}$

33. Simplify each fraction.

   a. $\dfrac{8x}{2}$ b. $\dfrac{3x}{6}$ c. $\dfrac{50}{5x}$ d. $\dfrac{6y^2}{12y}$

34. Simplify each expression. Write your result as a single fraction.

   a. $\dfrac{9x^6}{18x^2}$ b. $\dfrac{14x^7y^4}{12x^3y^3}$ c. $\dfrac{14a^4b^3}{35a^4b^3}$ d. $\dfrac{6x^A}{4x^B}$

35. The expression $\dfrac{8^X}{8^Y}$ is equivalent to 8^M if $M =$ _____.

36. The previous scenarios illustrate <u>The Quotient Rule</u>. It applies whenever you divide <u>like</u> bases. Write out the rule in your own words as if you were explaining it to someone.

37. Use the Quotient Rule to simplify each expression.

   a. $\dfrac{x^{12}}{x^4}$ b. $\dfrac{14f^{10}g^{15}}{7f^5g^8}$ c. $\dfrac{10x^{10}y^2z^3}{12x^7y^2z}$

NOTES

Use this page to record important ideas in the previous section or for any other writing that helps you learn the topics in this book.

Section 4
RAISING AN EXPONENT TO AN EXPONENT

GUIDED DISCOVERY SCENARIOS

38. Fill in the blanks to write the expression in an expanded form. The first one is done for you.

   a. $\left(2^2\right)^3 = 2^2 \cdot 2^2 \cdot 2^2$ b. $\left(2^3\right)^4 =$ _____   c. $\left(2^{10}\right)^5 =$ _____

39. How many 2's are multiplied together to form each expression?

   a. $\left(2^2\right)^3$ b. $\left(2^3\right)^4$ c. $\left(2^5\right)^{10}$ d. $\left(2^8\right)^x$

40. How many x's are multiplied together to form each expression?

   a. $\left(x^4\right)^2$ b. $\left(x^2\right)^5$ c. $\left(x^6\right)^{11}$ d. $\left(x^4\right)^Y$

41. Rewrite the expression $\left(B^{10}\right)^3$ as a base raised to an exponent.

42. Write each expression below as a repeatedly multiplied expression.

   a. $(2x)^3$ b. $(-3x)^3$

43. Try to simplify each expression. In this case, "simplify" means to write an equivalent form of the expression shown without parentheses in your expression.

   a. $\left(5x^4\right)^2$ b. $(-2y)^3$ c. $\left(\dfrac{z^2}{3}\right)^4$

44. How would you write $\left(-3x^{10}\right)^3$ as a repeated multiplication expression?

45. Simplify each expression.

   a. $\left(10x^7\right)^2$ b. $\left(-4y^2\right)^3$ c. $\left(-5xy^2\right)^4$

GUIDED DISCOVERY SCENARIOS

46. Simplify each expression.

   a. $\left(\dfrac{x}{3}\right)^4$

   b. $\left(\dfrac{5}{x^2}\right)^2$

   c. $\left(\dfrac{-2a^2}{b^3}\right)^3$

47. Simplify each expression as much as you can.

   a. $x^2 \cdot (2x)^3$

   b. $(5x)^2 \cdot (-2x)^2$

   c. $3(2x)^3 \cdot (3x)^2$

48. Determine the missing exponent(s) in each scenario.

   a. $(a^M)^N = a^?$

   b. $(xy)^N = x^? y^?$

   c. $\left(\dfrac{a}{b}\right)^K = \dfrac{a^?}{b^?}$

49. Determine the missing exponents in each scenario.

   a. $(x^L y^M)^N = x^? y^?$

   b. $\left(\dfrac{a^G}{b^H}\right)^K = \dfrac{a^?}{b^?}$

50. These scenarios illustrate <u>The Power Rule</u>. It applies when you raise an expression with an exponent to another exponent (if you raise a power to a power). Write The _____ Rule in your own words.

51. ★The value of H is $2x$ and the value of $H + 1$ is W. Prove that the value of $2^x \cdot 2^x + 2^H$ is 2^W.

52. Simplify each expression as much as you can.

   a. $2(2x+3x)^2$

   b. $-3(4y-y)^3$

NOTES

Use this page to record important ideas in the previous section or for any other writing that helps you learn the topics in this book.

Section 5
EXPONENTS REVIEW

GUIDED DISCOVERY SCENARIOS

53. The scenarios so far have used only positive exponents. This topic gets more difficult when the exponents have negative values. Use a calculator to find the value of each expression below. Write each result as a fraction.

   a. 2^{-1}
   b. 10^{-1}
   c. 5^{-1}

54. Use a calculator to find the value of each expression below.

   a. 2^0
   b. 10^0
   c. 5^0

55. Fill in the chart below. Use a calculator for the negative exponents. When results contain decimals, write them as fractions only. Use the chart below to learn about negative exponents and the exponent of 0.

   $2^4 =$ $(-3)^4 =$ $4^4 =$

   $2^3 =$ $(-3)^3 =$ $4^3 =$

   $2^2 =$ $(-3)^2 =$ $4^2 =$

   $2^1 =$ $(-3)^1 =$ $4^1 =$

   $2^0 =$ $(-3)^0 =$ $4^0 =$

   $2^{-1} =$ $(-3)^{-1} =$ $4^{-1} =$

   $2^{-2} =$ $(-3)^{-2} =$ $4^{-2} =$

   $2^{-3} =$ $(-3)^{-3} =$ $4^{-3} =$

   $2^{-4} =$ $(-3)^{-4} =$ $4^{-4} =$

56. Without a calculator, guess the value of each expression below.

   a. 5^{-1}
   b. 6^{-1}
   c. 7^{-2}
   d. 8^0

GUIDED DISCOVERY SCENARIOS

57. At this point, you should know a little bit more about negative exponents and the effect of an exponent of 0. You may feel ready to move on, but first, fill in the next chart. This chart is designed to help you become more familiar with evaluating a fraction that is raised to an exponent.

$(-½)^4 =$ $(⅓)^4 =$ $(-¼)^4 =$

$(-½)^3 =$ $(⅓)^3 =$ $(-¼)^3 =$

$(-½)^2 =$ $(⅓)^2 =$ $(-¼)^2 =$

$(-½)^1 =$ $(⅓)^1 =$ $(-¼)^1 =$

$(-½)^0 =$ $(⅓)^0 =$ $(-¼)^0 =$

$(-½)^{-1} =$ $(⅓)^{-1} =$ $(-¼)^{-1} =$

$(-½)^{-2} =$ $(⅓)^{-2} =$ $(-¼)^{-2} =$

$(-½)^{-3} =$ $(⅓)^{-3} =$ $(-¼)^{-3} =$

$(-½)^{-4} =$ $(⅓)^{-4} =$ $(-¼)^{-4} =$

58. Without a calculator, guess the value of each expression below.

   a. $\left(\dfrac{1}{5}\right)^{-1}$ b. $\left(\dfrac{1}{6}\right)^{-1}$ c. $\left(\dfrac{2}{3}\right)^{-2}$ d. $\left(\dfrac{5}{7}\right)^{0}$

NOTES

Use this page to record important ideas in the previous section or for any other writing that helps you learn the topics in this book.

Section 6
THE EXPONENT OF ZERO

GUIDED DISCOVERY SCENARIOS

59. Use the patterns that develop in the previous scenarios to determine the value of each expression.

   a. 5^0
   b. $(-2)^0$
   c. $\left(\dfrac{1}{3}\right)^0$
   d. x^0

60. Consider the expression $2^0 x^0$. It is clear that the 2 and the x both have exponents of 0 because those exponents are visible. Now consider the expression $2x^5$.

   a. What is the exponent of the x?

   b. What is the exponent of the 2?

   c. In the expression $9x^0$, what is the exponent of the 9?

   d. What is the value of $67m^0$?

61. What is the value of each expression below?

   a. $3(2)^0$
   b. $-2(-5)^0$
   c. $7x^0$
   d. $(7x)^0$

62. Why do the expressions $7x^0$ and $(7x)^0$ have different values?

63. Simplify each expression.

   a. $\dfrac{6x^6}{2x^2}$
   b. $\dfrac{8y^8}{6y^6}$
   c. $\dfrac{10x^4 y^7}{4x^3 y^5}$

64. Simplify each expression as much as you can.

   a. $3x^3 \cdot 2x^2$
   b. $(3x)^3 \cdot (2x)^2$
   c. $(-3x)^3 \cdot (-3x)^2$

Section 7
NEGATIVE EXPONENTS

GUIDED DISCOVERY SCENARIOS

65. Recall the patterns you noticed in the previous scenarios to determine the value of each of the following expressions. Express your answer as a fraction.

    a. 6^{-1}
    b. 7^{-1}
    c. 8^{-1}
    d. X^{-1}

66. In a previous scenario, you found that $2^2 = 4$ while $2^{-2} = \frac{1}{4}$. Additionally, $2^3 = 8$ while $2^{-3} = \frac{1}{8}$.

    a. What is the relationship between these pairs of results?

    b. Since $2^4 = 16$, what is the value of 2^{-4}?

    c. What is the value of 4^{-3}?

67. How can you find the value of A^X if A is a positive integer and X is a negative integer?

68. You have seen earlier that 3^{-1} has the same value as $\left(\frac{1}{3}\right)^1$. Since 3 can be written as $\frac{3}{1}$, it follows that $\left(\frac{3}{1}\right)^{-1} = \left(\frac{1}{3}\right)^1$. Following this structure, what is the value of $\left(\frac{1}{3}\right)^{-1}$?

69. Determine the value of each of the following expressions.

    a. $\left(\frac{2}{3}\right)^{-1}$
    b. $\left(\frac{4}{5}\right)^{-1}$
    c. $\left(\frac{A}{B}\right)^{-1}$

70. Determine the value of each of the following expressions.

    a. $\left(\frac{6}{7}\right)^{-2}$
    b. $\left(\frac{9}{8}\right)^{-2}$
    c. $\left(\frac{A}{B}\right)^{-2}$

71. What is the value of 0^{-1}?

GUIDED DISCOVERY SCENARIOS

72. What is the value of x^{-1}?

73. What is the value of x^{-3}?

74. Since $4^{-2} = \dfrac{1}{4^2}$ and $y^{-2} = \dfrac{1}{y^2}$, how would you write $(4y)^{-2}$ using only positive exponents?

75. In the expression $3xy^2$, what is the exponent of the 3? What is the exponent of the x?

76. In the expression $2x^{-6}$, what is the exponent of the 2? What is the exponent of the x?

77. Rewrite $4y^{-2}$ using only positive exponents.

78. Why is the expression $4y^{-2}$ different than $(4y)^{-2}$?

79. In the following examples, there are negative exponents. Rewrite each expression using only positive exponents.

   a. $2x^{-3}$ b. $6^{-2}x^2$ c. x^2y^{-3} d. $\dfrac{1}{y^{-4}}$ e. $\dfrac{x}{y^{-2}}$

Now that you know a little bit more about negative exponents, let's go back through a selection of the exponent scenarios that you worked on in previous scenarios.

80. Simplify each expression using only positive exponents in your final answer.

   a. $x^{-1} \cdot x^{-5}$ b. $7p^{-2}$ c. $\dfrac{3x^3}{x^{-7}}$

GUIDED DISCOVERY SCENARIOS

81. Simplify each expression using only positive exponents in your final answer.

    a. $\left(2^{-2}\right)^3$
    b. $\left(g^3\right)^{-4}$
    c. $\left(3x\right)^{-4}$

82. What is the most simplified form of $\left(\dfrac{3}{4}\right)^{-2}$?

83. Describe how to determine the value of a fraction that is raised to a negative exponent. Create a specific example using a fraction and an exponent of your choice.

84. Simplify each expression using only positive exponents.

    a. $\left(-\dfrac{3}{4}\right)^{-2}$
    b. $\left(\dfrac{x}{4}\right)^{-2}$
    c. $\left(\dfrac{3}{x}\right)^{-2}$
    d. $\left(-\dfrac{x}{y}\right)^{-2}$

85. Simplify the expression $\left(-\dfrac{xy^{-5}}{2}\right)^{-2}$ using only positive exponents.

86. Try to simplify each expression. Include only positive exponents in your final result.

    a. $\left(\dfrac{x^3}{3}\right)^{-4}$
    b. $\left(\dfrac{5}{x^2}\right)^{-2}$
    c. $\left(3y^{-2}\right)^3$

GUIDED DISCOVERY SCENARIOS

87. Simplify the expression $\left(-5x^{-2}y^3\right)^{-3}$ using only positive exponents.

88. Simplify the expression $\left(\dfrac{2^{-1}a^5}{b^{-2}}\right)^{-3}$ using only positive exponents.

89. Simplify each expression using only positive exponents.

   a. $\dfrac{x^4}{x^{12}}$

   b. $\dfrac{7f^5g^{10}}{14f^9g^8}$

   c. $\dfrac{12x^7y^2z}{10x^{10}y^2z^3}$

90. Which expressions are equivalent to $\dfrac{1}{3^4}$?

   a. $-1 \cdot 3^4$
   b. $3^1 \cdot 3^3$
   c. $3^{-2} \cdot 3^{-2}$
   d. $3^{-4} \cdot 3^0$
   e. $3^{-7} \cdot 3^3$

91. What is the value of 0^{-2}?

92. What is the value of $(3x)^{-2}$?

93. Rewrite each expression using only <u>negative</u> exponents. Yes, negative exponents. Think about what you have been doing to convert negative exponents into positive exponents and reverse that process.

   a. $\dfrac{1}{x^2}$
   b. $\dfrac{1}{x}$
   c. $\dfrac{3}{y^4}$
   d. $x^{-3} \cdot x^{-1}$
   e. $\dfrac{y^{-2}}{y^5}$

NOTES

Use this page to record important ideas in the previous section or for any other writing that helps you learn the topics in this book.

Section 8
EVALUATING EXPRESSIONS WITH EXPONENTS

GUIDED DISCOVERY SCENARIOS

94. The expression 5x has no specific value assigned to it until you replace x with a number. If x is 2, then the value of the expression 5x is 5(2) or 10.

   a. If x is −3, then the value of 5x is 5(−3), which equals _____, when simplified.

   b. If x is −1, what is the value of 5x?

   c. If x is −1, what is the value of $5x^4$?

95. Evaluate each expression if x is replaced with −2. Do NOT use a calculator.

   a. −x
   b. x^2
   c. $-x^2$

96. Evaluate each expression if x is replaced with −3. Do NOT use a calculator.

   a. $(-x)^2$
   b. $\frac{1}{3}x^3$
   c. $-2x^3$

97. Evaluate the expression $\frac{-x^3}{2}$, if x is given the value shown below.

   a. x = −1
   b. x = −2
   c. x = −3

98. Without a calculator, evaluate each expression if x is assigned the value of −3 and y is assigned the value of −2. Do NOT use a calculator.

   a. $\frac{y^2}{x^3}$
   b. $x^3 - 2x^2 + 3x$
   c. $-\frac{3}{4}y^2 + \frac{1}{2}y$

GUIDED DISCOVERY SCENARIOS

99. Evaluate each expression if $x = 3.7$ and $y = 1.4$. Calculators ARE allowed here. Round answers to 3 decimal places.

   a. $\dfrac{y^2}{x^3}$

   b. $0.03x^3 - 5.7x^2 + 9.4x$

   c. $-0.13y^2 + 0.5y$

100. Without a calculator, evaluate the expression $5x^{-2}$ if $x = 2$.

101. ★Without a calculator, evaluate the expression $x^{-2} - x^{-3}$ if $x = \dfrac{1}{2}$.

102. Explain why 2^{-1} and -2 have different numerical values.

103. Simplify each expression shown.

   a. $\left(\dfrac{2}{3}\right)^{-3} \div \left(\dfrac{1}{3}\right)^{-2}$

   b. $\dfrac{5x^{-4}}{xy^{-2}}$

   c. $\left(3x^{-2}\right)^{-3}$

104. Simplify each expression shown.

   a. $-3x^2 \cdot 4x \cdot 2x^{-5}$

   b. $3 - 2(-1)^2$

   c. $\left(-5x^7\right)^2$

NOTES

Use this page to record important ideas in the previous section or for any other writing that helps you learn the topics in this book.

Section 9
CUMULATIVE REVIEW

105. Write the equation for the line shown, using the Slope-Intercept Form.

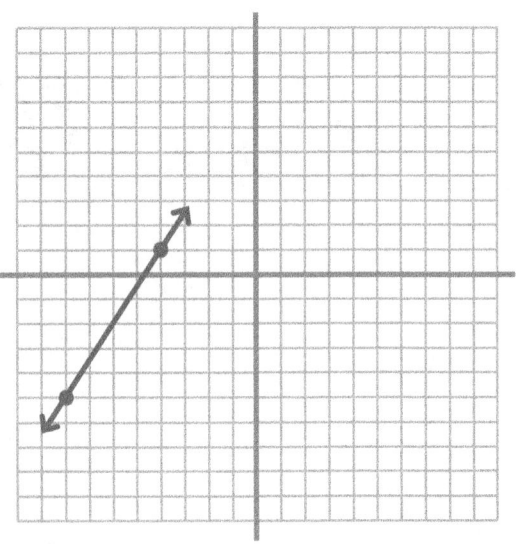

106. Draw a line that is perpendicular to the line in the previous scenario. The line that you draw must pass through the ordered pair $(3,-1)$.

107. Identify the equation of the line that you drew in the previous scenario.

108. The average price of a home is $360,000 in the urban section close to the center of a city. In the rural section outside the city limits, the average price of a home is $240,000.

   a. By what percent does the average price of a home change as you move from the urban section to the rural section?

   b. By what percent does the average price of a home change as you move from the rural section to the urban section?

GUIDED DISCOVERY SCENARIOS

109. You planted an apple tree, but it takes a few years for a tree to grow apples. Last year, the tree finally started producing fruit and you picked 12 apples. This year, your apple harvest increased by 150%. How many apples did you pick this year?

110. One variety of a strawberry plant will only produce fruit for a few years. Last year was a good year for your strawberry plant, but this year, the plant produced 25% fewer strawberries. If you picked 390 strawberries this year, how many did you pick last year?

111. You arrive at school early and realize that your laptop has a low battery so you find an open outlet and start charging the battery. Even though it is unrealistic, suppose that the battery charges at a constant rate until it is completely charged. When you plug in the laptop at 7:20am, the device is 24% charged. At 7:36am, you check the progress and find that the laptop is 48% charged.

   a. Suppose school starts at 8:00am. Will it be fully charged?

   b. By what percent is the charge increasing every minute?

112. Bea and Jo take separate walks every morning. If Bea travels $\frac{2}{3}$ mile every $\frac{1}{6}$ hour and Jo travels $\frac{7}{5}$ mile every 20 minutes, who walks at a slower pace?

113. Rewrite each expression using only positive exponents.

   a. $3x^{-2}$

   b. $3^{-2}x^3$

   c. $\left(\frac{1}{2}\right)^{-1}$

   d. $\left(\frac{1}{2}x\right)^{-3}$

NOTES

Use this page to record important ideas in the previous section or for any other writing that helps you learn the topics in this book.

Section 10
ANSWER KEY

1.	192 squares
2.	a. 2 b. 8 c. 32
3.	a. 6 b. 8
4.	2^N
5.	2^{10}
6.	a. 5^3 b. 10^4 c. $(-3)^5$
7.	a. x^3 b. y^4 c. z^5
8.	a. $10 \cdot 10 \cdot 10$ b. $(-2)(-2)(-2)(-2)$ c. $m \cdot m \cdot m \cdot m \cdot m$
9.	a. $3 \cdot 3 \cdot 3 \cdot 3 \cdot 3$ b. $y \cdot y \cdot y \cdot y \cdot y \cdot y$
10.	$2x \cdot 2x \cdot 2x \cdot 2x$
11.	$(x+1)(x+1)$
12.	a. 3^5 b. 4^8 c. 5^9 d. 6^{20}
13.	a. 3^7 b. $(-1)^{10}$ c. 2^{x+y}
14.	a. x^3 b. x^5 c. x^8
15.	a. x^7 b. y^{10} c. x^{A+B+C}
16.	When multiplying like bases, you can add the exponents
17.	a. $5^2 \cdot 5^3 = 5^5$ b. $3^5 \cdot 3^4 = 3^9$
18.	a. $10^2 \cdot x^2$ b. $4^3 y^2$
19.	a. $2^6 6^{10}$ b. $7^3 x^6$ c. $m^9 p^4$
20.	a. 3^{2x} b. 8^{2y+5} c. 5^{3x+1}
21.	a. 15 b. x^{10} c. $2x^7$
22.	a. $-4x^3$ b. $21y^{10}$ c. $45x^9$
23.	a. $3w^5 \cdot 3w^5 \to 9w^{10}$ b. $18x^6$ c. $-4m^7 \cdot -4m^7 \to 16m^{14}$
24.	$\frac{2}{1}$ or just 2
25.	two times
26.	2^3
27.	a. two b. three c. six
28.	a. 3^2 b. 7^2 c. 4^2
29.	a. 3^2 b. 7^2 c. 9^1 or just "9"
30.	a. 2^3 b. 4^1 or just "4" c. 5^0 or just "1"
31.	a. 5^3 b. $(-1)^3$ c. x^3
32.	a. x b. x^3 c. 1 d. x^8 e. x^{A-B}
33.	a. $4x$ b. $\frac{x}{2}$ c. $\frac{10}{x}$ d. $\frac{y}{2}$
34.	a. $\frac{x^4}{2}$ b. $\frac{7x^4 y}{6}$ c. $\frac{2}{5}$ d. $\frac{3x^{A-B}}{2}$
35.	$X - Y$
36.	When dividing like bases, simplify the fraction formed by the coefficients and subtract the exponents.
37.	a. x^8 b. $2f^5 g^7$ c. $\frac{5x^3 z^2}{6}$
38.	b. $2^3 \cdot 2^3 \cdot 2^3 \cdot 2^3$ c. $2^{10} \cdot 2^{10} \cdot 2^{10} \cdot 2^{10} \cdot 2^{10}$
39.	a. 6 b. 12 c. 50 d. $8X$
40.	a. 8 b. 10 c. 66 d. $4Y$
41.	$B^{10} \cdot B^{10} \cdot B^{10} \to B^{30}$
42.	a. $(2x)(2x)(2x)$ b. $(-3x)(-3x)(-3x)$
43.	a. $25x^8$ b. $-8y^3$ c. $\frac{z^8}{81}$
44.	$(-3x^{10})(-3x^{10})(-3x^{10})$
45.	a. $100x^{14}$ b. $-64y^6$ c. $625x^4 y^8$
46.	a. $\frac{x^4}{81}$ b. $\frac{25}{x^4}$ c. $\frac{-8a^6}{b^9}$
47.	a. $8x^5$ b. $100x^4$ c. $216x^5$
48.	a. a^{MN} b. $x^N y^N$ c. $\frac{a^K}{b^K}$
49.	a. $x^{LN} y^{MN}$ b. $\frac{a^{GK}}{b^{HK}}$
50.	Power; When raising a power to a power, multiply the exponents.
51.	$2^{2x} + 2^H = 2^H + 2^H = 2 \cdot 2^H = 2^{1+H} = 2^W$
52.	a. $2(5x)^2 \to 2 \cdot 25x^2 \to 50x^2$ b. $-3(3y)^3 \to -3 \cdot 27y^3 \to -81y^3$
53.	a. $\frac{1}{2}$ b. $\frac{1}{10}$ c. $\frac{1}{5}$
54.	a. 1 b. 1 c. 1

55.	$2^4 = 16$ $(-3)^4 = 81$ $4^4 = 256$ $2^3 = 8$ $(-3)^3 = -27$ $4^3 = 64$ $2^2 = 4$ $(-3)^2 = 9$ $4^2 = 16$ $2^1 = 2$ $(-3)^1 = -3$ $4^1 = 4$ $2^0 = 1$ $(-3)^0 = 1$ $4^0 = 1$ $2^{-1} = \frac{1}{2}$ $(-3)^{-1} = -\frac{1}{3}$ $4^{-1} = \frac{1}{4}$ $2^{-2} = \frac{1}{4}$ $(-3)^{-2} = \frac{1}{9}$ $4^{-2} = \frac{1}{16}$ $2^{-3} = \frac{1}{8}$ $(-3)^{-3} = -\frac{1}{27}$ $4^{-3} = \frac{1}{64}$ $2^{-4} = \frac{1}{16}$ $(-3)^{-4} = \frac{1}{81}$ $4^{-4} = \frac{1}{256}$
56.	a. $\frac{1}{5}$ b. $\frac{1}{6}$ c. $\frac{1}{7^2} \rightarrow \frac{1}{49}$ d. 1
57.	$\frac{1}{16}$ $\frac{1}{81}$ $\frac{1}{256}$ $-\frac{1}{8}$ $\frac{1}{27}$ $-\frac{1}{64}$ $\frac{1}{4}$ $\frac{1}{9}$ $\frac{1}{16}$ $-\frac{1}{2}$ $\frac{1}{3}$ $-\frac{1}{4}$ 1 1 1 -2 3 -4 4 9 16 -8 27 -64 16 81 256
58.	a. 5 b. 6 c. $\left(\frac{3}{2}\right)^2 \rightarrow \frac{9}{4}$ d. 1
59.	a. 1 b. 1 c. 1 d. 1
60.	a. 5 b. 1 c. 1 d. 67
61.	a. 3 b. -2 c. 7 d. 1
62.	In the expression $7x^0$, the 7 has an exponent of 1. In the expression $(7x)^0$, the parentheses make the 7 have an exponent of 0.
63.	a. $3x^4$ b. $\frac{4y^2}{3}$ c. $\frac{5xy^2}{2}$
64.	a. $6x^5$ b. $108x^5$ c. $-243x^5$
65.	a. $\frac{1}{6}$ b. $\frac{1}{7}$ c. $\frac{1}{8}$ d. $\frac{1}{X}$
66.	a. They are reciprocals b. $\frac{1}{16}$ c. $\frac{1}{64}$
67.	Evaluate the value of A^X as if X was positive and then write the reciprocal of your result.
68.	$\frac{3}{1}$ or 3
69.	a. $\frac{3}{2}$ b. $\frac{5}{4}$ c. $\frac{B}{A}$
70.	a. $\frac{49}{36}$ b. $\frac{64}{81}$ c. $\frac{B^2}{A^2}$
71.	$\frac{1}{0}$, which is undefined
72.	$\frac{1}{x}$
73.	$\frac{1}{x^3}$
74.	$\frac{1}{4^2 y^2}$ or $\frac{1}{16y^2}$
75.	3 has an exponent of 1 x has an exponent of 1
76.	2 has an exponent of 1 x has an exponent of -6
77.	$4 \cdot \frac{1}{y^2}$ or $\frac{4}{y^2}$
78.	In the term $4y^{-2}$, the 4 is raised to the exponent of 1. In the term $(4y)^{-2}$, the 4 is raised to the exponent of -2.
79.	a. $\frac{2}{x^3}$ b. $\frac{x^2}{36}$ c. $\frac{x^2}{y^3}$ d. y^4 e. xy^2
80.	a. $\frac{1}{x^6}$ b. $\frac{7}{p^2}$ c. $3x^{10}$
81.	a. $\frac{1}{2^6}$ or $\frac{1}{64}$ b. $\frac{1}{g^{12}}$ c. $\frac{1}{3^4 x^4}$ or $\frac{1}{81x^4}$
82.	$\left(\frac{4}{3}\right)^2 \rightarrow \frac{16}{9}$
83.	Answers may vary
84.	a. $\frac{16}{9}$ b. $\frac{16}{x^2}$ c. $\frac{x^2}{9}$ d. $\frac{y^2}{x^2}$
85.	$\frac{4y^{10}}{x^2}$
86.	a. $\frac{81}{x^{12}}$ b. $\frac{x^4}{25}$ c. $\frac{27}{y^6}$
87.	$\frac{x^6}{-125y^9}$
88.	$\frac{8}{a^{15}b^6}$
89.	a. $\frac{1}{x^8}$ b. $\frac{g^2}{2f^4}$ c. $\frac{6}{5x^3 z^2}$
90.	c., d. and e.

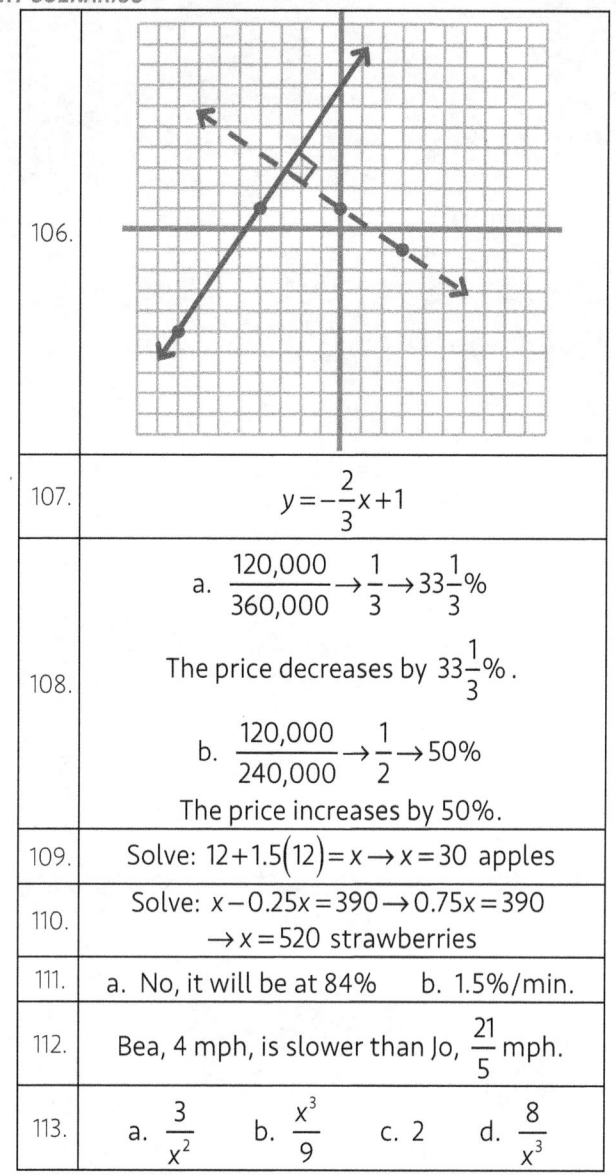

91.	$\frac{1}{0}$, which is undefined
92.	$\frac{1}{(3x)^2} = \frac{1}{9x^2}$
93.	a. x^{-2} b. x^{-1} c. $3y^{-4}$ d. x^{-4} e. y^{-7}
94.	a. −15 b. −5 c. 5
95.	a. 2 b. 4 c. −4
96.	a. 9 b. −9 c. 54
97.	a. 0.5 b. 4 c. 13.5
98.	a. $-\frac{4}{27}$ b. −54 c. −4
99.	a. 0.039 b. −41.733 c. 0.445
100.	$5x^{-2} = 5(2)^{-2} = \frac{5}{2^2} = \frac{5}{4}$
101.	$\left(\frac{1}{2}\right)^{-2} - \left(\frac{1}{2}\right)^{-3} = 2^2 - 2^3 = 4 - 8 = -4$
102.	$2^{-1} = \frac{1}{2} \to \frac{1}{2} \neq -2$ A negative exponent makes a number become its reciprocal. A negative sign makes a number become its opposite.
103.	a. $\left(\frac{3}{2}\right)^3 \div (3)^2 \to \frac{27}{8} \div 9 \to \frac{27}{8} \cdot \frac{1}{9} \to \frac{3}{8}$ b. $\frac{5y^2}{x^5}$ c. $\left(\frac{3}{x^2}\right)^{-3} \to \left(\frac{x^2}{3}\right)^3 \to \frac{x^6}{27}$
104.	a. $-24x^{-3} \to \frac{-24}{x^3}$ b. $3-2 \to 1$ c. $25x^{14}$
105.	$y = \frac{3}{2}x + 7$
106.	(graph showing two perpendicular lines)
107.	$y = -\frac{2}{3}x + 1$
108.	a. $\frac{120,000}{360,000} \to \frac{1}{3} \to 33\frac{1}{3}\%$ The price decreases by $33\frac{1}{3}\%$. b. $\frac{120,000}{240,000} \to \frac{1}{2} \to 50\%$ The price increases by 50%.
109.	Solve: $12 + 1.5(12) = x \to x = 30$ apples
110.	Solve: $x - 0.25x = 390 \to 0.75x = 390$ $\to x = 520$ strawberries
111.	a. No, it will be at 84% b. 1.5%/min.
112.	Bea, 4 mph, is slower than Jo, $\frac{21}{5}$ mph.
113.	a. $\frac{3}{x^2}$ b. $\frac{x^3}{9}$ c. 2 d. $\frac{8}{x^3}$

HOMEWORK & EXTRA PRACTICE SCENARIOS

As you complete scenarios in this part of the book, you will practice what you learned in the guided discovery sections. You will develop a greater proficiency with the vocabulary, symbols and concepts presented in this book. Practice will improve your ability to retain these ideas and skills over longer periods of time.

There is an Answer Key at the end of this part of the book. Check the Answer Key after every scenario to ensure that you are accurately practicing what you have learned. If you struggle to complete any scenarios, try to find someone who can guide you through them.

CONTENTS

Section 1 **REVIEW** **45**

Section 2 **INTRODUCTION TO EXPONENTS** **48**

Section 3 **MULTIPLYING EXPRESSIONS WITH EXPONENTS** **50**

Section 4 **DIVIDING EXPRESSIONS WITH EXPONENTS** **53**

Section 5 **RAISING AN EXPONENT TO AN EXPONENT** **56**

Section 6 **EXPONENTS REVIEW** **59**

Section 7 **THE EXPONENT OF ZERO** **61**

Section 8 **NEGATIVE EXPONENTS** **63**

Section 9 **EVALUATING EXPRESSIONS WITH EXPONENTS** **68**

Extra Review Sections

Section 10 **PRE-ALGEBRA REVIEW** **71**

Section 11 **BOOK 1 REVIEW** **74**

Section 12 **BOOK 2 REVIEW** **78**

Section 13 **BOOK 3 REVIEW** **83**

Section 14 **CUMULATIVE REVIEW** **87**

Section 15 **ANSWER KEY** **89**

Section 1
REVIEW

HOMEWORK & EXTRA PRACTICE SCENARIOS

1. Write the reciprocal of each expression below.

   a. $\dfrac{2}{3}$
   b. 5
   c. $\dfrac{1}{10}$
   d. $\dfrac{9x}{7y}$

2. Solve each equation.

   a. $\dfrac{2}{5}A = 8$
   b. $2B - 10B = 24$
   c. $C + 3(C - 2) = 34$

3. Solve each equation.

   a. $5x - 11 = -2 - 13x$
   b. $2x - 14 + 8x = 136$

4. Answer each question. Try to do this without a calculator.

   a. What percent of 40 is 60?
   b. 12 is 75% of what number?

5. A student answered 36 questions correctly on a test. She earned a grade of 90%. How many questions were on the test, if each question was worth the same number of points?

HOMEWORK & EXTRA PRACTICE SCENARIOS

6. Use the graph to answer the following question.

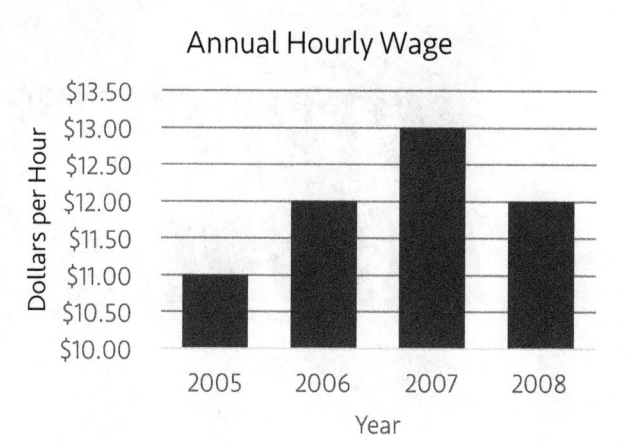

    a. Between which two consecutive years did the greatest percent of change occur?

    b. Between which two consecutive years did the smallest percent of change occur?

    Annual Hourly Wage
    (Bar graph: 2005 ≈ $11.00, 2006 ≈ $12.00, 2007 ≈ $13.00, 2008 ≈ $12.00; y-axis: Dollars per Hour; x-axis: Year)

7. In the previous scenario, the hourly wage in 2005 was 10% higher than in 2004. What was the hourly wage in 2004?

8. Circle the equations that form a line.

    a. $2x+3y=7$
    b. $y=2x^2+3$
    c. $2x^2+5y^2=10$
    ★d. $x^2+y=3+x^2$

9. Consider the line shown.

    a. Find the equation that shows the relationship between R and n.

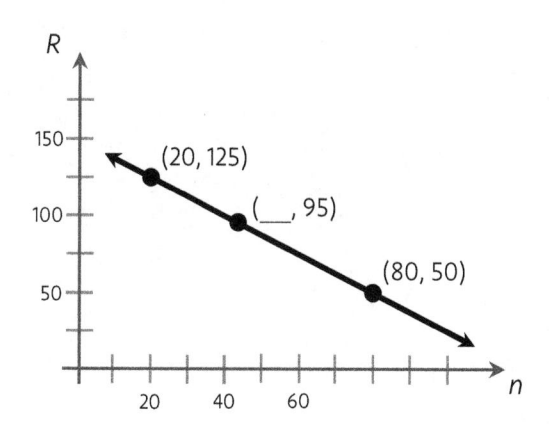

    (Graph showing line through points (20, 125), (__, 95), and (80, 50); axes R and n)

    b. Determine the missing value in the graph.

10. Simplify each expression as much as you can.

    a. $0.10-2.05$
    b. one-half of $\frac{1}{3}$
    c. $\left(\frac{3}{10}\right)^2$

Section 2
INTRODUCTION TO EXPONENTS

11. Write each of the following expressions as a single number raised to an exponent.

    a. $0 \cdot 0 \cdot 0$
    b. $1 \cdot 1 \cdot 1 \cdot 1$
    c. $(-2) \cdot (-2) \cdot (-2) \cdot (-2) \cdot (-2)$

12. Write each of the following expressions as a single variable raised to an exponent.

    a. $A \cdot A \cdot A \cdot A$
    b. $B \cdot B \cdot B$
    c. $C \cdot C$

13. Write each of the following expressions as a repeated multiplication expression. For example, 2^3 can be written as $2 \cdot 2 \cdot 2$.

    a. $(-1)^4$
    b. $5y^3$
    c. $-x^4$
    d. $4^2 \cdot 4^3$
    e. $3x^3 \cdot x \cdot x^2$

14. How would you write $(5y)^5$ as a repeated multiplication expression?

15. How would you write $(y-2)^3$ as a repeated multiplication expression?

16. Which number has a larger value?

    a. 2^3 or 3^2
    b. 2^4 or 4^2
    c. 3^4 or 4^3

17. Which number has a smaller value?

    a. $\left(\dfrac{1}{2}\right)^2$ or $\left(\dfrac{1}{2}\right)^3$
    b. $\left(\dfrac{4}{5}\right)^{20}$ or $\left(\dfrac{11}{10}\right)^2$

18. Which number is greater, option A or option B?

    Option A: $\left(\dfrac{1{,}332}{1{,}320}\right)^{27}$
    Option B: $\left(\dfrac{3{,}561}{3{,}572}\right)^{29}$

Section 3
MULTIPLYING EXPRESSIONS WITH EXPONENTS

HOMEWORK & EXTRA PRACTICE SCENARIOS

19. Consider the result when expressions containing exponents are multiplied together. Simplify each product below and write each result as a single number raised to an exponent.

   a. $1^2 \cdot 1^3$
   b. $2^3 \cdot 2^5$
   c. $3^1 \cdot 3^5$
   d. $4 \cdot 4^4$

20. Write each expression as a single number raised to an exponent.

   a. $10 \cdot 10^7 \cdot 10^2$
   b. $(-2)^7 \cdot (-2) \cdot (-2)^2$
   c. $5^A \cdot 5^B \cdot 5^C$

21. Write each expression as a single variable raised to an exponent.

   a. $y^3 \cdot y$
   b. $y^3 \cdot y^3$
   c. $y^3 \cdot y^7$

22. Write each expression as a single variable raised to an exponent.

   a. $A \cdot A^2 \cdot A^8$
   b. $B^{10} \cdot B \cdot B^3$
   c. $C^x \cdot C^y \cdot C^z$

23. Write an explanation of The Product Rule, which applies when you multiply like bases and determine the exponent of your result.

24. Is $2^3 \cdot 2^3$ the same thing as 4^6 or does $2^3 \cdot 2^3$ equal 2^6? How can you persuade someone else that your conclusion is accurate?

25. Simplify each expression to make each base appear only once.

   a. $3^2 \cdot 3^2 \cdot 4^7 \cdot 4$
   b. $5 \cdot 5 \cdot y \cdot y$
   c. $f \cdot g^2 \cdot g^5 \cdot f$

26. Simplify each expression to make each base appear only once.

   a. $9m \cdot 9m \cdot 9m \cdot 9m$

   b. $13x^3 \cdot 13x^3 \cdot 13x^3$

27. Simplify each expression to make each base appear only once.

   a. $7^y \cdot 7^y$

   b. $10^{30} \cdot 10$

   b. $11^A \cdot 11 \cdot 11^3$

28. Fill in the blank.

   a. $6x^4 \cdot 2x^4 = \underline{} x^8$

   b. $8x^3 \cdot x = 8\underline{}$

   c. $-5x^2 \cdot 6x^8 = \underline{}$

29. Simplify each expression as much as you can.

   a. $-y^3 \cdot 5y$

   b. $6x^2 \cdot 3x^5$

   c. $3y^4 \cdot -2y^2 \cdot y$

30. Simplify each expression.

   a. $(-7w^2)(-4w^7)$

   b. $2x^{10} \cdot 9x^{-4}$

   c. $(8h^{-3}) \cdot (2.5h^{15})$

31. Simplify each expression.

   a. $(5x^3)^2$

   b. $(-9y^5)^2$

Section 4
DIVIDING EXPRESSIONS WITH EXPONENTS

Now consider the result when expressions containing exponents are expressed in fractions.

32. How many times does a disguised form of 1 appear in each expression below?

    a. $\dfrac{2\cdot 2\cdot 2\cdot 2\cdot 2}{2\cdot 2\cdot 2\cdot 2}$

    b. $\dfrac{6\cdot 6\cdot 6\cdot 6}{6}$

    c. $\dfrac{10^7}{10\cdot 10\cdot 10\cdot 10\cdot 10}$

33. Write each expression in the previous scenario as a single number raised to an exponent.

34. Write each expression as a single number raised to an exponent.

    a. $\dfrac{9^{15}}{9^{14}}$

    b. $\dfrac{6^5}{6^5}$

    c. $\dfrac{1^7}{1^3}$

35. Write each expression as a single number raised to an exponent.

    a. $12^7 \div 12$

    b. $20^{17} \div 20^{10}$

    c. $y^{100} \div y^{70}$

36. The expression $\dfrac{2^A}{2^B}$ is equivalent to 2^C if $C =$ _____.

37. Write each expression as a single variable raised to an exponent.

    a. $\dfrac{x^3}{x}$

    b. $\dfrac{x^8}{x^3}$

    c. $\dfrac{x^3}{x^3}$

    d. $\dfrac{x^{12}}{x^6}$

    e. $\dfrac{x^C}{x^D}$

38. Simplify each fraction.

    a. $\dfrac{2}{8x}$

    b. $\dfrac{6}{3x}$

    c. $\dfrac{5x}{50}$

    d. $\dfrac{12y}{6y^2}$

HOMEWORK & EXTRA PRACTICE SCENARIOS

39. Write an explanation of The Quotient Rule.

40. Use the Quotient Rule to simplify each expression.

   a. $\dfrac{25f^7}{5f^4}$

   b. $\dfrac{12x^2y^5}{18x^2y}$

41. Simplify each expression as much as you can.

   a. $\dfrac{3p^{12}g^{40}}{12p^{10}g^{39}}$

   b. $\dfrac{16a^X}{18a^Y}$

42. Simplify each expression.

   a. $\dfrac{x^6}{x}$

   b. $\dfrac{4a^9b^{13}}{9b^8a^9}$

   c. $\dfrac{16x^5y^8z^3}{12xy^3z^3}$

43. Simplify each expression.

   a. $\dfrac{5(2y)^6}{16y^5}$

   b. $\dfrac{9(2x)^2}{(3x)^2}$

44. If $x = 3$, what is the value of $x^2 - 5x + 1$?

Section 5
RAISING AN EXPONENT TO AN EXPONENT

45. Fill in the blanks to write the expression in an expanded form. The first one is done for you.

   a. $\left(3^7\right)^2 = 3^7 \cdot 3^7$

   b. $\left(3^2\right)^7 = $ _____

46. How many 3's are multiplied together to form each expression?

   a. $\left(3^7\right)^2$ b. $\left(3^2\right)^7$ c. $\left(3^4\right)^5$ d. $\left(3^P\right)^7$

47. How many y's are multiplied together to form each expression?

   a. $\left(y^4\right)^6$ b. $\left(y^3\right)^8$ c. $\left(y^2\right)^{12}$ d. $\left(y^A\right)^B$

48. How would you write $\left(-2y^5\right)^3$ as a repeated multiplication expression?

49. How would you write $\left(5xy^2\right)^4$ as a repeated multiplication expression?

50. Try to simplify each expression. In this case, "simplify" means to write an equivalent form of the expression shown without parentheses in your expression. Take your time with these and avoid the temptation to make quick assumptions.

   a. $(-x)^3$ b. $(-3y)^3$ c. $\left(\dfrac{z^4}{2}\right)^5$

51. Simplify each expression.

   a. $\left(3x^5\right)^3$ b. $\left(-y^3\right)^5$ c. $\left(-10x^3y\right)^2$

52. Simplify each expression.

   a. $\left(-\dfrac{x}{2}\right)^3$
   b. $\left(\dfrac{2}{x^3}\right)^4$
   c. $\left(\dfrac{-2a^2}{b^7}\right)^5$

53. Simplify each expression as much as you can.

   a. $x^4 \cdot (3x)^4$
   b. $(2x^2)^2 \cdot (-4x)^2$
   c. $-2(-5x)^2 \cdot (-2x)^3$

54. Determine the missing exponents in each scenario.

   a. $\left(x^A y^B\right)^C = x^? y^?$
   b. $\left(\dfrac{x^A}{y^B}\right)^C = \dfrac{x^?}{y^?}$

55. Write out an explanation of The Power Rule.

56. Simplify each expression as much as you can.

   a. $5(12y - 2y)^3$
   b. $-(x - 5x)^2$

Section 6
EXPONENTS REVIEW

57. Fill in the blanks shown. When results contain decimals, write them as <u>fractions</u>. Use the chart below to examine the effect of negative exponents and the exponent of 0.

$(-2)^3 =$ $\qquad$ $10^3 =$ $\qquad$ $\left(\frac{2}{3}\right)^3 =$

$(-2)^2 =$ $\qquad$ $10^2 =$ $\qquad$ $\left(\frac{2}{3}\right)^2 =$

$(-2)^1 =$ $\qquad$ $10^1 =$ $\qquad$ $\left(\frac{2}{3}\right)^1 =$

$(-2)^0 =$ $\qquad$ $10^0 =$ $\qquad$ $\left(\frac{2}{3}\right)^0 =$

$(-2)^{-1} =$ $\qquad$ $10^{-1} =$ $\qquad$ $\left(\frac{2}{3}\right)^{-1} =$

$(-2)^{-2} =$ $\qquad$ $10^{-2} =$ $\qquad$ $\left(\frac{2}{3}\right)^{-2} =$

$(-2)^{-3} =$ $\qquad$ $10^{-3} =$ $\qquad$ $\left(\frac{2}{3}\right)^{-3} =$

58. Simplify each expression as much as you can.

   a. $-2y^4 \cdot 3y^2 \cdot y$
   b. $2^4 \cdot 2^2$
   c. $\dfrac{9x^2 y^4}{15x^3 y}$

59. Simplify each expression as much as you can.

   a. $\left(\dfrac{-2y^2}{x^3}\right)^3$
   b. $(x^2)^3 + (x^3)^2$
   c. $-3(-y)^2 \cdot (-2y)^3$

60. Without using a calculator, which expression has a larger value?

   Expression 1: $(3^2)^3$ $\qquad$ Expression 2: $3^2 + 3^2 + 3^2 + 3^2$

Section 7
THE EXPONENT OF ZERO

61. Determine the value of each expression.

    a. 2^0
    b. $(-7)^0$
    c. $\left(\dfrac{3}{5}\right)^0$
    d. y^0

62. Consider the expression $5^1 y^1$. It is clear that the 5 and the y both have exponents of 1 because those exponents are visible. Now consider the expression $7p$.

    a. What is the exponent of the 7?

    b. What is the exponent of the p?

    c. In the expression $3x^2$, what is the exponent of the 3?

    d. What is the value of $3x^0$?

    e. What is the value of $-2x^0$?

63. What is the value of each expression below?

    a. $7(9)^0$
    b. $-(-8)^0$
    c. $-5y^0$
    d. $(-5y)^0$

64. Why do the expressions $-5y^0$ and $(-5y)^0$ have different values?

65. Simplify each expression as much as you can.

    a. $5(-3f+f)^3$
    b. $-(g-2g)^4 + 2g^0$

Section 8
NEGATIVE EXPONENTS

HOMEWORK & EXTRA PRACTICE SCENARIOS

66. Use the patterns that you noticed in earlier scenarios to determine the value of each the following expressions. Express your answer as a fraction.

   a. 6^{-1} b. 1^{-1} c. $(-2)^{-1}$ d. y^{-1}

67. Consider the following: $10^1 = 10$ while $10^{-1} = \dfrac{1}{10}$. Also, $\left(\dfrac{2}{3}\right)^2 = \dfrac{4}{9}$ while $\left(\dfrac{2}{3}\right)^{-2} = \dfrac{9}{4}$.

   a. What is the relationship between these pairs of results?

   b. Since $3^4 = 81$, what is the value of 3^{-4}?

   c. What is the value of 5^{-2}?

68. How can you determine the value of x^E if x is an integer and E is negative?

69. What is the value of $\left(\dfrac{1}{5}\right)^{-1}$?

70. Determine the value of each the following expressions.

   a. $\left(\dfrac{1}{9}\right)^{-1}$ b. $\left(\dfrac{6}{7}\right)^{-1}$ c. $\left(\dfrac{x}{y}\right)^{-1}$

71. Determine the value of each the following expressions.

   a. $\left(\dfrac{9}{2}\right)^{-2}$ b. $\left(\dfrac{3}{10}\right)^{-2}$ c. $\left(\dfrac{x}{y}\right)^{-2}$

72. What is the value of 0^{-3}? Why is this?

73. Since $3^{-5} = \dfrac{1}{3^5}$ and $x^{-5} = \dfrac{1}{x^5}$, how would you write $(3x)^{-5}$ using only positive exponents?

74. In the expression $-3x^2$, what is the exponent of the -3? What is the exponent of the x?

75. Rewrite $2x^{-4}$ using only positive exponents.

76. Why is the expression $2x^{-4}$ different than $(2x)^{-4}$?

77. Rewrite each expression using only positive exponents.

   a. $9y^{-2}$ b. $2^{-4}x^4$ c. $x^{-5}y^2$ d. $\dfrac{1}{x^{-1}}$ e. $\dfrac{x}{y^{-1}}$

Now that you know a little bit more about negative exponents, let's go back through a selection of the exponent scenarios that you worked on in previous scenarios.

78. Simplify each expression using only positive exponents in your final answer.

   a. $y^{-3} \cdot y^{-1}$ b. $2z^{-5}$ c. $\dfrac{6x^2}{x^{-4}}$

79. Simplify each expression using only positive exponents in your final answer.

   a. $(3^{-1})^2$ b. $(f^8)^{-2}$ c. $(10g)^{-2}$

80. Describe how to determine the value of a fraction that is raised to a negative exponent. Create a specific example using a fraction and an exponent of your choice.

81. Write each expression in two ways, first as a single variable raised to a negative exponent, and then as a fraction using only positive exponents.

   a. $\dfrac{x}{x^3}$
   b. $\dfrac{x^3}{x^8}$
   c. $\dfrac{x^6}{x^{12}}$
   d. $\dfrac{x^5}{x^{15}}$

82. Try to simplify each expression using only positive exponents.

   a. $\left(-5x^{-1}\right)^2$
   b. $\left(-\dfrac{y}{4}\right)^{-2}$
   c. $\left(-\dfrac{7}{p^5}\right)^{-3}$
   d. $\left(-9x^6y^{-1}\right)^{-2}$

83. Simplify the expression $\left(\dfrac{8^{-2}x^3}{y^{-7}}\right)^{-3}$ using only positive exponents.

84. Simplify each expression. Write your result as a single fraction, using only positive exponents.

   a. $\dfrac{5f^4}{25f^7}$
   b. $\dfrac{18x^2y}{12x^2y^5}$
   c. $\dfrac{12p^{10}g^{39}}{3p^{12}g^{40}}$

HOMEWORK & EXTRA PRACTICE SCENARIOS

85. Simplify the expression $\dfrac{18a^Y}{16a^X}$.

86. Simplify each expression.

   a. $\dfrac{x}{x^6}$

   b. $\dfrac{9a^8 b^9}{4b^9 a^{13}}$

   c. $\dfrac{12xy^3 z^3}{16x^5 y^8 z^3}$

87. Rewrite each expression using only <u>negative</u> exponents. Yes, negative exponents. Think about what you have been doing to convert negative exponents into positive exponents and reverse that process.

   a. $\dfrac{1}{5}$

   b. $\dfrac{2}{x^2}$

   c. $\left(\dfrac{1}{3}\right)^3$

   d. $x^3 \cdot x^{-1}$

   e. $\dfrac{x^3}{x^{-1}}$

88. Simplify each expression below.

   a. $(-4)^2$

   b. -4^2

   c. $-5(-1)^3$

   d. $5 - 3(-2) - 4(-2)^2$

89. Simplify each expression shown.

   a. $\dfrac{1}{4}(3x)^2 \cdot (-2x^2)^3$

   b. $\dfrac{9f^{11} m^{-2}}{27 f^5 m^{-7}}$

Section 9
EVALUATING EXPRESSIONS WITH EXPONENTS

90. In the expression $-y$, the negative sign represents a –1. In the expression $-y^2$, the negative sign also represents a –1.

    a. What is the value of $-2y^2$ if $y = 1$?

    b. What is the value of $-2y^2$ if $y = -1$?

    c. What is the value of $-y^2$ if $y = 1$?

    d. What is the value of $-y^2$ if $y = -1$?

91. Evaluate each expression if y is assigned the value of –3. Calculators are NOT allowed here.

    a. $-2y$

    b. $3y^2$

    c. $-4y^3$

92. Evaluate each expression if H is assigned the value of –10. Calculators are NOT allowed here.

    a. $(-H)^4$

    b. $\dfrac{1}{100}H^5$

93. Evaluate each expression if x is assigned the value of –1 and y is assigned the value of –4. Calculators are NOT allowed here.

    a. $\dfrac{y^2}{x^7}$

    b. $-x^4 + 7x^3 - 2x^2$

    c. $-\dfrac{3}{4}y^3 + \dfrac{1}{8}y^2$

94. Evaluate each expression if x is assigned the value of 1.5 and y is assigned the value of 2.5. Calculators ARE allowed here. Round answers to 3 decimal places.

   a. $\dfrac{x^5}{y^2}$

   b. $-1.5y^3 + 3.9y^2 - 2.1y$

   c. $4.7x^2 + 9.2x$

95. Without a calculator, evaluate the expression $x^{-1} + x^{-2}$ if $x = -2$.

96. ★Without a calculator, evaluate the expression $x^{-1} + x^{-2} + x^{-3}$ if $x = -\dfrac{1}{2}$.

97. Fill in the blanks below to show what you have learned so far about properties of exponents.

   a. $A^x A^y = $ _____

   b. $\dfrac{A^y}{A^x} = $ _____

   c. $\left(A^x\right)^y = $ _____

   d. $A^0 = $ _____

   e. $AB^0 = $ _____

   f. $A^{-1} = $ _____

   g. $AB^{-1} = $ _____

   h. $\left(\dfrac{A}{B}\right)^{-1} = $ _____

Section 10
PRE-ALGEBRA REVIEW

Extra Review Section

The following scenarios involve the operations of addition, subtraction, multiplication, division, fractions and simple equations.

98. Without a calculator, which expression is larger? Support your choice with calculations.

   Expression #1: $77 + 14$ Expression #2: $113 - 22$

99. Without a calculator, which expression is larger? Support your choice with calculations.

   Expression #1: $\dfrac{2}{3} + \dfrac{5}{4}$ Expression #2: $\dfrac{5}{2} - \dfrac{3}{5}$

100. Without a calculator, which expression is larger? Support your choice with calculations.

   Expression #1: $\dfrac{2}{3} \cdot \dfrac{6}{7}$ Expression #2: $\dfrac{3}{15} \div \dfrac{2}{5}$

101. Some equations can be solved using one step, or one operation. Consider the equations below. Describe the <u>one operation</u> you can perform to solve each equation. Do not solve the equation.

   a. $x - 0.5 = 7.8$ b. $\dfrac{x}{3} = 1.5$ c. $\dfrac{2}{3}x = 8$

102. In each equation in the previous scenario, is the value of x bigger or smaller than the number on the right side of the equation? Do not solve the equation to answer this. Instead, make a logical guess.

103. Solve each equation in the previous scenario.

104. Solve each equation.

    a. $0.4a + a = 28$

    b. $4 - \frac{1}{3}b = 9$

105. Solve each equation.

    a. $-4x + 5 = 17 - 2x$

    b. $\frac{60 + 45 + c}{3} = 40$

Section 11
BOOK 1 REVIEW

EXTRA REVIEW SECTION

In Book 1, you learned how to calculate percent changes and how to write and solve equations to represent percentage scenarios.

106. Without a calculator, which expression has a larger value? Explain your choice with words.

    Expression #1: 49% of 4   Expression #2: 26% of 8

107. Write and solve an equation to answer each question.

    a. What number is 75% of 32?   b. 18% of 910 is what number?

108. Write and solve an equation to answer each question.

    a. 40% of what number is 20?   b. 113 is 3.1% of what number?

109. Write and solve an equation to answer each question.

    a. 30 is what percent of 50?   b. What percent of 16 is 54?

110. On Monday, 90 students were sick and had to stay home from school. On Tuesday, 135 students were sick. By what percent did the number of sick students increase from Monday to Tuesday?

111. The whale population dropped from 315,000 to 290,000. By what percent did the whale population change?

112. The value of a savings account was $1,000 on Tuesday. The value of the account decreased by 5% from Tuesday to Wednesday. What was the value of the account on Wednesday?

113. You have some of your money in a bank account. After the bank realizes it made a mistake, it increases your account by 5%. Your account is now worth $1,000. How much money was in your account before the bank realized it made a mistake?

114. The population of a small town decreases by 10% in one year to 9,000 people.

   a. How many people were in the town before the decrease?

   b. A common response is to increase 9000 by 10% to find the population last year. Why does this not calculate the original population?

115. The final bill for a new TV is $638.40 after tax is calculated. If the sales tax is 6.4%, what was the original cost of the TV?

At the end of Book 1, you learned about rates.

116. What rate can be calculated in the chart shown?

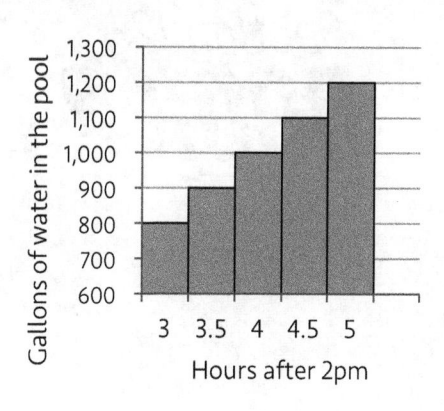

117. The pool in the previous scenario is being filled with water. How much water was in the pool at 2pm?

118. In the previous scenario, how much water will be in the pool at 9pm?

119. In October, a Monarch butterfly begins its southward migration to Mexico. After 8 days of flying, the butterfly is spotted 1200 miles away from its eventual destination. After 12 days of flying, the butterfly is 900 miles away from its destination.

   a. How far will the Monarch have traveled by the time it reaches its destination?

   b. How many total days will it take for the Monarch to arrive at its destination in Mexico?

120. Suppose $T = 18 - 2A$.

   a. What is the value of T if $A = 9$?

   b. What is the value of A if $T = 11$?

Section 12
BOOK 2 REVIEW

EXTRA REVIEW SECTION

HOMEWORK & EXTRA PRACTICE SCENARIOS

121. The Smith family owns an apple orchard and they sell their apples. The cost of the apples depends on the total weight of the apples.

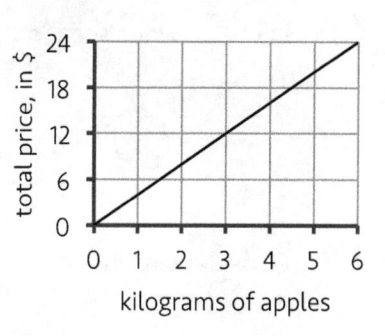

    a. What is the rate shown in the graph? Identify the numerical value of the rate and express the rate using proper units.

    b. Write an equation that shows the total price, P, if someone buys k kilograms of apples.

122. The rate in the previous graph can be found by calculating the slope of the line. Determine the slope of the line shown to the right.

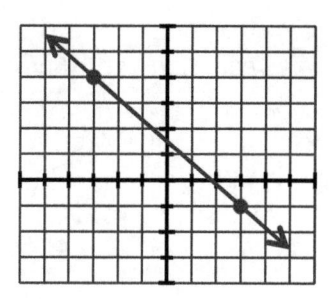

123. The slope in the previous scenario can be measured by using the graph. Sometimes, you will not have a graph. Suppose a line passes through the points $(-25, 11)$ and $(-10, 6)$. Without graphing these points, what is the slope of this line?

124. Graph the lines on the same Cartesian Plane.

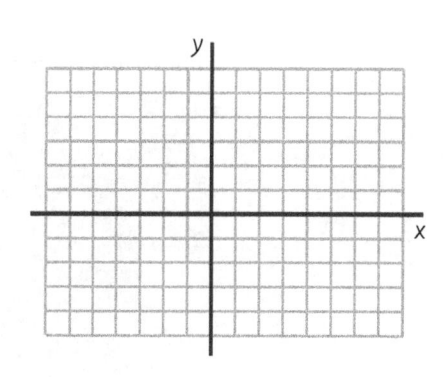

    a. $y = \dfrac{3}{4}x - 2$

    b. $y = -\dfrac{4}{3}x + 3$

    c. Describe what you notice when you look at the way that these lines appear on the graph.

HOMEWORK & EXTRA PRACTICE SCENARIOS

125. In the previous scenario, the two lines are exactly perpendicular. Why does this occur?

126. Consider the line shown in the graph. Identify the equation of this line.

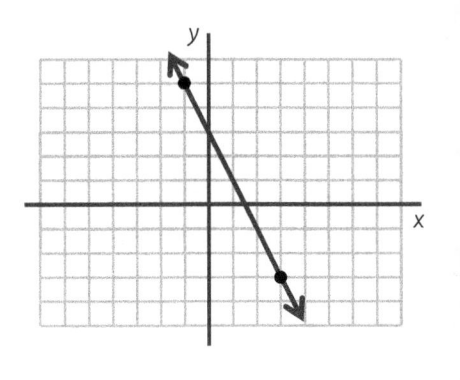

127. If a point is located on a line, it "satisfies" the equation of that line. In the previous scenario, the equation of the line is $y = -2x + 3$. Do any of the points below satisfy the equation?

    a. (2, –1)   b. (–1, 4)   c. (15, –28)

128. Identify the y-intercept of the line shown.

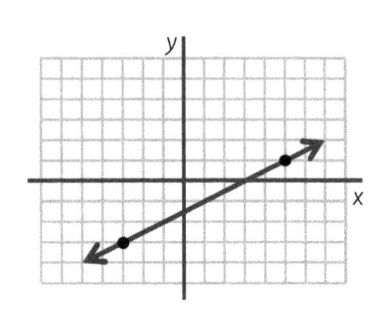

129. A line has a slope of $-\dfrac{2}{5}$. It passes through the point $(10, 8)$. What is the y-intercept of the line?

130. Consider the ordered pairs $(-5, 1)$ and $(5, -3)$. Determine the equation of the line that passes through the given ordered pairs and then graph the line.

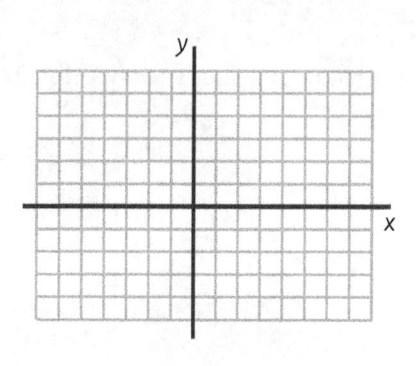

131. What is the equation of the line that passes through the points $(-1, -4)$ and $(-6, 11)$?

132. If an ordered pair has an x-value of 0, where will that point be located on the Cartesian plane?

133. If an ordered pair has a y-value of 0, where will it be located on the Cartesian plane?

134. Consider the graph shown to the right.

   a. Identify the x-intercept of the line shown to the right.

   b. Identify the y-intercept of the line shown to the right.

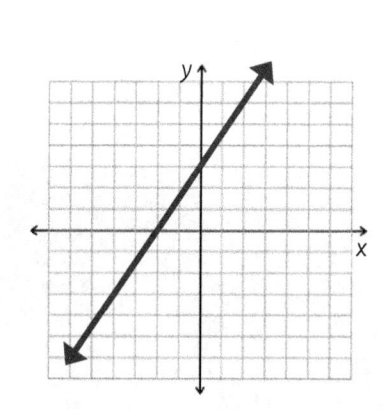

135. Consider the equation 2x + 5y = –10.

   a. Find the coordinates of the x- and y-intercepts of the equation.

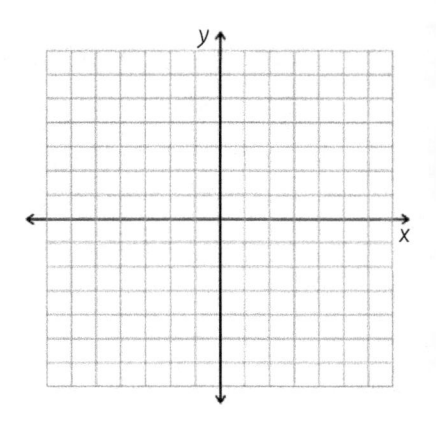

   b. Find the coordinates of one more point on the line and then graph the line.

136. Two lines are shown on the Cartesian Plane to the right.

   a. What is the slope of the dashed line?

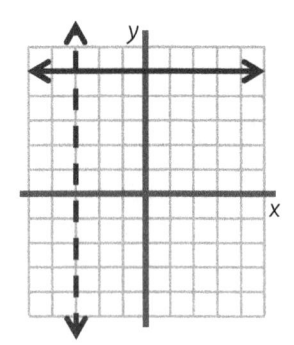

   b. What is the slope of the solid line?

Section 13
BOOK 3 REVIEW

EXTRA REVIEW SECTION

137. Simplify each expression as much as you can.

   a. $(-10)^3$

   b. $5-1(-3)^3$

   c. $(7^2-5\cdot 9)^2 \div (-1)^9$

138. Simplify each expression as much as you can.

   a. $11-5(-2)^3$

   b. $4^2-(3^2-1)^2 \div (-2)^5$

139. Simplify each expression as much as you can.

   a. $x \cdot x^3 \cdot x^5$

   b. $\dfrac{y^{70}}{y^{10}}$

   c. $x+x$

140. Simplify each expression as much as you can, using only positive exponents in your answer.

   a. $2x \cdot 3x^2$

   b. x^{-2}

   c. $\dfrac{2x^{-4}}{3y^{-1}}$

141. Simplify each expression as much as you can.

    a. $-2y^4 \cdot 3y^2 \cdot y$

    b. $2^4 \cdot 2^2$

    c. $\dfrac{9x^2y^4}{15x^3y}$

142. Simplify each expression as much as you can.

    a. $\left(\dfrac{-2y^2}{x^3}\right)^3$

    b. $(x^2)^3 + (x^3)^2$

    c. $-3(-y)^2 \cdot (-2y)^3$

143. Rewrite each expression using only positive exponents.

    a. $3x^{-2}$

    b. $3^{-2}x^3$

    c. $\left(\dfrac{1}{2}\right)^{-1}$

144. Rewrite each expression using only positive exponents.

    a. $y^0 + x^0 + 3w^0$

    b. $(x^{-2})^2$

    c. $\left(\dfrac{1}{2}x\right)^{-3}$

145. Without a calculator, evaluate each expression if $x = -1$ and $y = -2$.

   a. $x^2 + 3x - 7$

   b. $-2y^2 - 7y + 9$

146. Without a calculator, evaluate the expression $x^{-1} + x^{-2}$ if $x = -2$.

Section 14
CUMULATIVE REVIEW

147. Write the equation for the line shown, in Slope-Intercept Form.

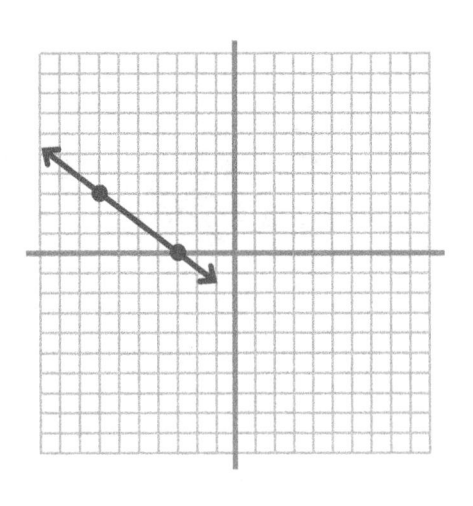

148. Draw a line that is perpendicular to the previous line and passes through (–3, 4).

149. What is the equation of the line you drew in the previous scenario?

150. The price of a cheese pizza costs $9.75, but you can add as many toppings as you want for $0.50 each. How many toppings are on a pizza that costs $20.25?

151. The average amount of money spent on Halloween per family is shown in the graph. Each mark on the vertical axis represents an increase of $5, but the lowest mark does not represent $0. If the average amount of money spent decreased by 12.5% from 2012 to 2013, what was the average amount spent per family in 2012?

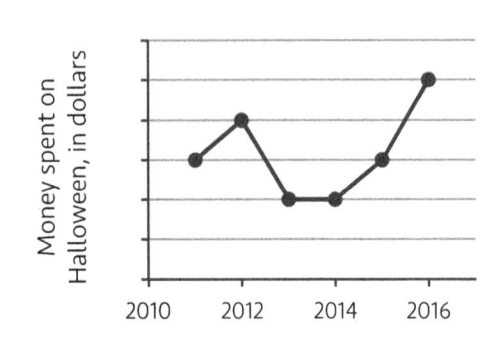

152. A recipe for 4 loaves of bread requires $\frac{5}{8}$ of an ounce of yeast. How much yeast is needed to bake two loaves of bread? Express the amount as a fraction.

Section 15
ANSWER KEY

1.	a. $\frac{3}{2}$ b. $\frac{1}{5}$ c. 10 d. $\frac{7y}{9x}$
2.	a. $\frac{5}{2} \cdot \frac{2}{5} A = \frac{8}{1} \cdot \frac{5}{2} \rightarrow A = 20$ b. $-8B = 24 \rightarrow B = -3$ c. $C + 3C - 6 = 34 \rightarrow 4C = 40 \rightarrow C = 10$
3.	a. $18x - 11 = -2 \rightarrow 18x = 9 \rightarrow x = 0.5$ b. $10x - 14 = 136 \rightarrow 10x = 150 \rightarrow x = 15$
4.	a. $x \cdot 40 = 60 \rightarrow x = 1.5 \rightarrow x = 150\%$ b. $12 = 0.75 \cdot x \rightarrow \frac{12}{0.75} = x \rightarrow x = 16$
5.	36 is 90% of the questions $36 = 0.9 \cdot Q \rightarrow \frac{36}{0.9} = Q \rightarrow Q = 40$
6.	a. 2005-2006 ($\approx 9.09\%$) b. 2007-2008 ($\approx 7.69\%$)
7.	In 2004, the hourly was $10. Solve $1.1x = 11 \rightarrow x = \$10$
8.	Circle a. and d. A line's equation can be written as $y = mx + b$ or as $Ax + By = C$.
9.	a. $R = -1.25n + 150$ b. 44
10.	a. -1.95 b. $\frac{1}{2} \cdot \frac{1}{3} \rightarrow \frac{1}{6}$ c. $\frac{9}{100}$
11.	a. 0^3 b. 1^4 c. $(-2)^5$
12.	a. A^4 b. B^3 c. C^2
13.	a. $(-1) \cdot (-1) \cdot (-1) \cdot (-1)$ b. $5 \cdot y \cdot y \cdot y$ c. $-1 \cdot x \cdot x \cdot x \cdot x$ d. $4 \cdot 4 \cdot 4 \cdot 4 \cdot 4$ e. $3 \cdot x \cdot x \cdot x \cdot x \cdot x \cdot x$
14.	$5y \cdot 5y \cdot 5y \cdot 5y \cdot 5y$
15.	$(y-2)(y-2)(y-2)$
16.	a. 3^2 ($9 > 8$) b. both expressions = 16 c. 3^4 ($81 > 64$)
17.	a. $\left(\frac{1}{2}\right)^2 \rightarrow \frac{1}{4} > \frac{1}{8}$ b. $\left(\frac{4}{5}\right)^{20} \rightarrow$ A fraction slightly less than 1 gets closer to 0 as it is multiplied by itself repeatedly.
18.	Option A is larger. In Option A, the fraction is bigger than 1. With a positive exponent, it gets larger. In Option B, the fraction is less than 1. With a positive exponent, it gets smaller.
19.	a. 1^5 b. 2^8 c. 3^6 d. 4^5
20.	a. 10^{10} b. $(-2)^{10}$ c. 5^{A+B+C}
21.	a. y^4 b. y^6 c. y^{10}
22.	a. A^{11} b. B^{14} c. C^{x+y+z}
23.	When multiplying like bases, add the exponents: $4^5 \cdot 4^7 = 4^{12}$
24.	$2^6 \rightarrow$ Write $2^3 \cdot 2^3$ as $2 \cdot 2 \cdot 2 \cdot 2 \cdot 2 \cdot 2$.
25.	a. $3^4 \cdot 4^8$ b. $5^2 \cdot y^2$ c. $f^2 \cdot g^7$
26.	a. $(9m)^4$ b. $13^3 x^9$
27.	a. 7^{2y} b. $10^{30} \cdot 10$ c. 11^{4+A}
28.	a. 12 b. x^4 c. $-30x^{10}$
29.	a. $-5y^4$ b. $18x^7$ c. $-6y^7$
30.	a. $28w^9$ b. $18x^6$ c. $20h^{12}$
31.	a. $5x^3 \cdot 5x^3 \rightarrow 25x^6$ b. $-9y^5 \cdot -9y^5 \rightarrow 81y^{10}$
32.	a. 4 b. 1 c. 5
33.	a. 2^1 b. 6^3 c. 10^2
34.	a. 9^1 b. 6^0 or 1 c. 1^4 or 1
35.	a. 12^6 b. 20^7 c. y^{30}
36.	$A - B$
37.	a. x^2 b. x^5 c. 1 d. x^6 e. x^{C-D}
38.	a. $\frac{1}{4x}$ b. $\frac{2}{x}$ c. $\frac{x}{10}$ d. $\frac{2}{y}$
39.	When dividing like bases, simplify the fraction formed by the coefficients and subtract the exponents.
40.	a. $5f^3$ b. $\frac{2y^4}{3}$
41.	a. $\frac{p^2 g}{4}$ b. $\frac{8a^{X-Y}}{9}$
42.	a. x^5 b. $\frac{4b^5}{9}$ c. $\frac{4x^4 y^5}{3}$
43.	a. $\frac{5}{1} \cdot \frac{64y^6}{16y^5} \rightarrow 5 \cdot 4y \rightarrow 20y$

	b. $\dfrac{9}{1} \cdot \dfrac{4x^2}{9x^2} \to \dfrac{9}{1} \cdot \dfrac{4}{9} \to 4$
44.	$(3)^2 - 5(3) + 1 \to 9 - 15 + 1 \to -6 + 1 \to -5$
45.	b. $3^2 \cdot 3^2 \cdot 3^2 \cdot 3^2 \cdot 3^2 \cdot 3^2 \cdot 3^2$
46.	a. 14 b. 14 c. 20 d. 7P
47.	a. 24 b. 24 c. 24 d. AB
48.	$(-2y^5)(-2y^5)(-2y^5)$
49.	$(5xy^2)(5xy^2)(5xy^2)(5xy^2)$
50.	a. $-x^3$ b. $-27y^3$ c. $\dfrac{z^{20}}{32}$
51.	a. $27x^{15}$ b. $-y^{15}$ c. $100x^6 y^2$
52.	a. $-\dfrac{x^3}{8}$ b. $\dfrac{16}{x^{12}}$ c. $\dfrac{-32a^{10}}{b^{35}}$
53.	a. $81x^8$ b. $64x^6$ c. $400x^5$
54.	a. $x^{AC} y^{BC}$ b. $\dfrac{x^{AC}}{y^{BC}}$
55.	When raising a power to a power, multiply the exponents.
56.	a. $5(10y)^3 \to 5(1000y^3) \to 5000y^3$ b. $-(-4x)^2 \to -(16x^2) \to -16x^2$
57.	-8 1000 $\tfrac{8}{27}$ 4 100 $\tfrac{4}{9}$ -2 10 $\tfrac{2}{3}$ 1 1 1 $-\tfrac{1}{2}$ $\tfrac{1}{10}$ $\tfrac{3}{2}$ $\tfrac{1}{4}$ $\tfrac{1}{100}$ $\tfrac{9}{4}$ $-\tfrac{1}{8}$ $\tfrac{1}{1000}$ $\tfrac{27}{8}$
58.	a. $-6y^7$ b. 2^6 c. $\dfrac{3y^3}{5x}$
59.	a. $\dfrac{-8y^6}{x^9}$ b. $2x^6$ c. $24y^5$
60.	Expression 1 is larger Expression 1: $(3^2)^3 \to 9^3$ Expression 2: $9+9+9+9 \to 9 \cdot 4$
61.	a. 1 b. 1 c. 1 d. 1
62.	a. 1 b. 1 c. 1 d. 3 d. -2
63.	a. 7 b. -1 c. -5 d. 1
64.	In $-5y^0$, only the y has an exponent of 0. In $(-5y)^0$, the entire expression is raised to an exponent of 0.
65.	a. $5(-2f)^3 \to 5 \cdot -8f^3 \to -40f^3$

	b. $-(-g)^4 + 2 \cdot 1 \to -1 \cdot g^4 + 2 \to -g^4 + 2$
66.	a. $\tfrac{1}{6}$ b. 1 c. $-\tfrac{1}{2}$ d. $\tfrac{1}{y}$
67.	a. The results are reciprocals. b. $\tfrac{1}{81}$ c. $\tfrac{1}{25}$
68.	Evaluate the value of x^E as if E was positive and then write the reciprocal of your result.
69.	5
70.	a. 9 b. $\dfrac{7}{6}$ c. $\dfrac{y}{x}$
71.	a. $\dfrac{4}{81}$ b. $\dfrac{100}{9}$ c. $\dfrac{y^2}{x^2}$
72.	Undefined, because its value is $\dfrac{1}{0}$
73.	$\dfrac{1}{(3x)^5}$ or $\dfrac{1}{3^5 x^5}$
74.	-3 has an exponent of 1 x has an exponent of 2
75.	$\dfrac{2}{x^4}$
76.	In $2x^{-4}$, only the x has an exponent of -4. In $(2x)^{-4}$, the entire expression is raised to an exponent of -4.
77.	a. $\dfrac{9}{y^2}$ b. $\dfrac{x^4}{16}$ c. $\dfrac{y^2}{x^5}$ d. x e. xy
78.	a. $\dfrac{1}{y^4}$ b. $\dfrac{2}{z^5}$ c. $6x^6$
79.	a. $\dfrac{1}{9}$ b. $\dfrac{1}{f^{16}}$ c. $\dfrac{1}{100g^2}$
80.	Option 1: Evaluate the value of the fraction if the exponent is positive and then write the reciprocal of that result. Option 2: Write the reciprocal of the original fraction and then evaluate the value with a positive exponent.
81.	a. $x^{-2} \to \dfrac{1}{x^2}$ b. $x^{-5} \to \dfrac{1}{x^5}$ c. $x^{-6} \to \dfrac{1}{x^6}$ d. $x^{-10} \to \dfrac{1}{x^{10}}$
82.	a. $\dfrac{25}{x^2}$ b. $\dfrac{16}{y^2}$ c. $-\dfrac{p^{15}}{7^3}$ d. $\dfrac{y^2}{81x^{12}}$
83.	$\dfrac{8^6}{x^9 y^{21}}$

84.	a. $\dfrac{1}{5f^3}$ b. $\dfrac{3}{2y^4}$ c. $\dfrac{4}{p^2 g}$
85.	$\dfrac{9a^{y-x}}{8}$
86.	a. $\dfrac{1}{x^5}$ b. $\dfrac{9}{4a^5}$ c. $\dfrac{3}{4x^4 y^5}$
87.	a. 5^{-1} b. $2x^{-2}$ c. 3^{-3} d. $\dfrac{1}{x^{-2}}$ e. $\dfrac{1}{x^{-4}}$
88.	a. 16 b. −16 c. $-5 \cdot -1 = 5$ d. $5 + 6 - 4 \cdot 4 \to 11 - 16 \to -5$
89.	a. $\dfrac{1}{4} \cdot 9x^2 \cdot -8x^6 \to -18x^8$ b. $\dfrac{f^6 m^5}{3}$
90.	a. −2 b. −2 c. −1 d. −1
91.	a. 6 b. 27 c. 108
92.	a. $(-(-10))^4 \to 10^4 \to 10{,}000$ b. $\dfrac{1}{100}(-10)^5 \to \dfrac{-100{,}000}{100} \to -1{,}000$
93.	a. −16 b. −10 c. 50
94.	a. 1.215 b. −4.313 c. 24.375
95.	$(-2)^{-1} + (-2)^{-2} = \dfrac{1}{-2} + \dfrac{1}{(-2)^2} = -\dfrac{1}{2} + \dfrac{1}{4} = -\dfrac{1}{4}$
96.	$\left(-\dfrac{1}{2}\right)^{-1} + \left(-\dfrac{1}{2}\right)^{-2} + \left(-\dfrac{1}{2}\right)^{-3}$ $= (-2)^{-1} + (-2)^{-2} + (-2)^{-3} = -2 + 4 - 8 = -6$
97.	a. A^{x+y} b. A^{y-x} c. A^{xy} d. 1 e. $A(1) \to A$ f. $\dfrac{1}{A}$ g. $A \cdot \dfrac{1}{B} \to \dfrac{A}{B}$ h. $\dfrac{B}{A}$
98.	Both expressions are equal to 91.
99.	Expression #1 is larger. $\dfrac{8}{12} + \dfrac{15}{12} = \dfrac{23}{12} = 1\dfrac{11}{12}$ $\dfrac{25}{10} - \dfrac{6}{10} = \dfrac{19}{10} = 1\dfrac{9}{10}$ $\dfrac{11}{12}$ is larger than $\dfrac{9}{10}$.
100.	Expression #1 is larger. Expression #1: $\dfrac{12}{21} \to \dfrac{4}{7}$ Expression #2: $\dfrac{3}{15} \cdot \dfrac{5}{2} \to \dfrac{15}{30} \to \dfrac{1}{2}$
101.	a. Add 0.5 to both sides b. Multiply both sides by 3
	c. Multiply both sides by $\dfrac{3}{2}$
102.	a. x is greater than 7.8 (What number decreased by 0.5 equals 7.8?) b. x is greater than 1.5 (What number divided by 3 equals 1.5?) c. x is greater than 8 (Two-thirds of what number is 8?)
103.	a. $x = 8.3$ b. $x = 4.5$ c. $x = 12$
104.	a. $1.4a = 28 \to a = \dfrac{28}{1.4} \to a = 20$ b. $-\dfrac{1}{3}b = 5 \to b = -15$
105.	a. $5 = 17 + 2x \to -12 = 2x \to -6 = x$ b. $\dfrac{105 + c}{3} = 40 \to 105 + c = 120 \to c = 15$
106.	Expression #2 is larger. Since 49% is smaller than one-half, 49% of 4 is less than 2. Since 26% is larger than one-fourth, 26% of 8 is larger than 2 (one-fourth of 8 is 2).
107.	a. $n = 0.75(32) \to n = 24$ b. $0.18(910) = n \to n = 163.8$
108.	a. $0.4n = 20 \to n = 50$ b. $113 = 0.031n \to n \approx 3645.2$
109.	a. $30 = \dfrac{x}{100}(50) \to 30 = \dfrac{1}{2}x \to x = 60\%$ b. $\dfrac{x}{100}(16) = 54 \to \dfrac{4}{25}x = 54 \to x = 337.5\%$
110.	45 more students were sick on Tuesday. 45 is one-half of 90, or 50% of 90. The number of sick students increased by 50% from Monday to Tuesday.
111.	The population decreased about 7.9%. $315{,}000 - 290{,}000 = 25{,}000$ 25,000 is what percent of 315,000? $25{,}000 = x(315{,}000)$ $x = 7.94\%$ You can also calculate the percent by dividing the amount of change by the original amount.
112.	$950. Option 1: Find 95% of 1000 Option 2: Solve $1000 - 0.05(1000) = x$
113.	$952.38; solve $x + 0.05x = 1000$
114.	a. equation: $0.9P = 9{,}000 \to P = 10{,}000$ b. The original population decreases by 10%, which is not the same result as increasing the new population by 10%.
115.	$x + 0.064x = 638.40 \to 1.064x = 638.4$

	$x = 600 \to \$600$
116.	The amount of water in a pool is increasing by 100 gallons per half-hour, which is 200 gallons per hour.
117.	At 5pm, 800 gallons of water are in the pool. At 2pm (3 hours earlier) it has 200 gallons in it (600 fewer gallons).
118.	At 7pm, 1200 gallons of water are in the pool. At 9pm (2 hours later) it has 1600 gallons in it (400 more gallons).
119.	a. 1800 miles b. 24 days
120.	a. $T = 18 - 2(9) \to T = 0$ b. $11 = 18 - 2A \to -7 = -2A \to A = 3.5$
121.	a. $4 per kilogram b. $P = 4k$
122.	$-\dfrac{5}{6}$
123.	$\dfrac{6-11}{-10-(-25)} \to \dfrac{-5}{15} \to -\dfrac{1}{3}$
124.	a, b. [graph] c. The lines are perpendicular.
125.	The slopes of the two lines are opposite reciprocals.
126.	$y = -2x + 3$
127.	a. Yes, it is on the line b. No, it is not on the line c. No, when you replace x and y with 15 and −28, the equation becomes $-28 = -2(15) + 3$ or $-28 = -27$, which is a false statement.
128.	$(0, -1.5)$
129.	The y-intercept is (0, 12). In the equation $y = mx + b$, replace m with $-\dfrac{2}{5}$ and replace x and y with 10 and 8 to make the equation $8 = -\dfrac{2}{5}(10) + b$. Solve for b to get $b = 12$.
130.	$y = -\dfrac{2}{5}x - 1$
131.	[graph] $y = -3x - 7$
132.	on the y-axis
133.	on the x-axis
134.	a. x-int: $(-2, 0)$ b. y-int: $(0, 3)$
135.	a. To find the x-int, replace y with 0 in the equation and solve for x. To find the y-int, replace x with 0 in the equation and solve for y. x-int: $(-5, 0)$ y-int: $(0, -2)$ b. Graph:
136.	a. undefined b. 0
137.	a. -1000 b. $5 - 1(-27) \to 5 + 27 \to 32$ c. $(49-45)^2 \div (-1)^9 \to (4)^2 \div -1 \to -16$
138.	a. $11 - 5(-8) \to 11 + 40 \to 51$ b. $4^2 - (9-1)^2 \div (-2)^5 \to 16 - 8^2 \div (-32)$ $\to 16 - 64 \div (-32) \to 16 - (-2) \to 18$
139.	a. $x^{1+3+5} \to x^9$ b. $y^{70-10} \to y^{60}$ c. $2x$
140.	a. $6x^3$ b. $\dfrac{1}{x^2}$ c. $\dfrac{2y}{3x^4}$
141.	a. $-6y^7$ b. 2^6 c. $\dfrac{3y^3}{5x}$
142.	a. $\dfrac{-8y^6}{x^9}$ b. $x^6 + x^6 \to 2x^6$ c. $-3y^2 \cdot -8y^3 \to 24y^5$
143.	a. $\dfrac{3}{x^2}$ b. $\dfrac{x^3}{9}$ c. 2
144.	a. $1 + 1 + 3(1) \to 5$ b. $\dfrac{1}{x^4}$ c. $\dfrac{8}{x^3}$
145.	a. $(-1)^2 + 3(-1) - 7 \to 1 - 3 - 7 \to -2 - 7 = -9$ b. $-2(-2)^2 - 7(-2) + 9 \to -2(4) + 14 + 9 = 15$
146.	$(-2)^{-1} + (-2)^{-2} \to -\dfrac{1}{2} + \dfrac{1}{4} \to -\dfrac{1}{4}$

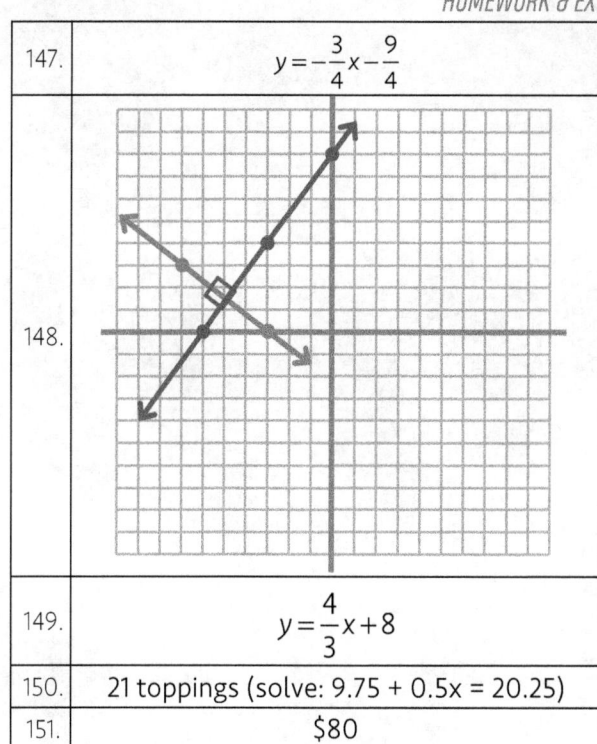

147.	$y = -\frac{3}{4}x - \frac{9}{4}$
148.	
149.	$y = \frac{4}{3}x + 8$
150.	21 toppings (solve: 9.75 + 0.5x = 20.25)
151.	$80

152.	one-half of $\frac{5}{8}$ → $\frac{2.5}{8}$ or $\frac{5}{16}$